Arrow B Ranch

Clemons Okeechobee, LLC

Matt Clemons

Carlton Ranch

Babcock Ranch

Bar G Ranch

Oscar Clemons

Harold McClelland

Syfrett Ranch

Barthle Brothers Ranch, Inc.

Bert & Mary Ann Tucker

Bill & Alice Ramsey

John Corbett

D.L. Thomas Ranch

Deseret Cattle & Citrus

Billy Peeples Ranch

Bronson Ranch

Buck Island Ranch

Dixie Ranch

Donald L. Plagge

Bobby R. Durden

Buck Mann Ranch

C Bar Ranch

Capote Cattle Co.

Ed Yarborough Ranches, Inc.

Farley Cattle Co.

Florida Commision Company

B.S. Cattle Company

Vieno Ranch

Adams Ranch

Brad Phares

Bill Phares

Beverly Hills Plantation, Inc.

XJ Cattle Company

Travis Larson

Thomas Cattle Company

Manufactured in the United States of America

Library of Congress Control Number: 2012956032

ISBN 978-0-9860337-0-4

Cover: Herding cattle through water, Horse Creek Ranch, Hardee County, 2012.
Photo by Carlton Ward, Jr.

L-R: Doyle Carlton III, Brian Alexy, and Dale Carlton move a few head of cattle with the help of yellow cur cow-dogs.

Opposite: Cattle on the beach. Apalachicola, 1900s. Courtesy of the State Archives of Florida, image n047144

FLORIDA CATTLE RANCHING

FIVE CENTURIES OF TRADITION

Florida Cattlemen's Foundation

The Florida Cattlemen's Foundation, a 501 (c) (3) corporation founded in 1999, is dedicated to the preservation of the history, heritage and ideals of Florida cattlemen and to the enhancement of beef production and marketing practices for future generations.

The Foundation's support of education, research, leadership programs will ensure the future and enhance the vitality of Florida beef industry. It will keep the rich heritage of cattle ranching viable and progressive in the years and generations to come through a variety of methods including:

- Research Funding
- Environmental Stewardship & Land Management Promotion
- Youth Educational Scholarships
- Preservation Historic Artifacts and Documents
- Youth Leadership Programs
- Donor Recognition
- Charitable Organization Support

To make a tax-exempt charitable donation, or to sponsor one of our programs or activities, please contact:

Florida Cattlemen's Foundation
Post Office Box 421929
Kissimmee, FL 34742-1929
www.floridacattlemen.org/foundation
407.846.6221

Exhibition Tour Dates

March 11-August 9, 2009 – Museum of Florida History, Tallahassee

September 19, 2009-December 19, 2009 – Tampa Bay History Center

January 18, 2010-July 24, 2010 – Western Folklife Center, Elko, NV

September 3, 2010-January 23, 2011 – HistoryMiami, Miami

February 12-May 8, 2011 – Florida Museum of Natural History, UF Gainesville

January 13-November 8, 2012 – South Florida Fair, West Palm Beach

February 7, 2013 – Long-term display at the Florida State Fair, Tampa

Foreword by Patrick Smith

There's more to Florida than beaches and theme parks. There's a whole different world out there, but you have to get off the main highways to find it. It's cattle country, and it probably gets fewer human footprints in one year than Disney World gets in five minutes.

Florida is the country's first cattle state—not the Southwest, or West, but here. There are still today, large and small, some 17,000 cattle ranches in Florida. One of them is the largest beef producer in the nation, and five are in the top ten.

Cattle ranches are Florida's protectors of the natural land and wildlife. Go far out on a ranch and you will see sandhill cranes and egrets and herons and grackles and doves and eagles and dozens more, as well as deer and raccoons and rabbits and squirrels, and maybe a native Florida panther.

Florida ranch country is still a place where families live together, work together, share together, laugh together, love together and die together—always with family first. It is the keeper of our Florida pioneer heritage, and our American heritage too.

Florida ranch country—may you last forever!

Patrick Smith is a 1999 inductee into the Florida Artists Hall of Fame. In May 2002, Smith was the recipient of the Florida Historical Society's Fay Schweim Award as the "Greatest Living Floridian" for the impact of his novels. In his honor, the Florida Historical Society created an annual award, the Patrick D. Smith Florida Literature Award. Best known for A Land Remembered, *the story of a pioneer family through three generations, Smith brings the Florida ranching heritage to life, vividly and powerfully—in a way that will never be forgotten. Learn more about* A Land Remembered *and his other prize-winning novels at www.patricksmithonline.com.*

FLORIDA DEPARTMENT OF AGRICULTURE AND CONSUMER SERVICES
COMMISSIONER ADAM H. PUTNAM
THE CAPITOL

November 28, 2012

Mr. Jim Strickland
Chairman
Florida Cattlemen's Foundation, Inc.
Post Office Box 421929
Kissimmee, Florida 34742-1929

Dear Mr. Strickland,

It has been a pleasure to work with you and your foundation to launch the Florida Cattle Ranching exhibit at the Florida State Fairgrounds. The Fairgrounds considers it a premier exhibit and I am confident it will be on every visitor's must see list.

Congratulations on the release of the outstanding feature book "Florida Cattle Ranching: Five Centuries of Tradition." The book is a stunning catalog depicting the cattle ranching exhibit and provides a historical account of this vital industry. I was especially impressed by the photos of the unique artifacts from the Spanish colonial period like the antiquated spurs and stirrups. I also enjoyed reading the progression from the industry's origins to modern day practices such as electronic identification tags.

The exhibit and book coincide with the statewide celebration of Florida's 500th birthday in 2013. It all started with the Spanish explorer, Juan Ponce de Leon's discovery of La Florida. Then in 1521, Ponce de Leon introduced cattle to the Florida landscape and gave birth to Florida's thriving cattle and ranching enterprise. Today, our state is home to five of the top ten largest cow/calf operations in the United States and boasts a national ranking of 10th in the number of beef cows. Nearly one-half of all Florida agricultural land is involved in cattle production and is home to countless wildlife and native plant system populations.

Please extend my appreciation to the Florida Cattlemen's Foundation and its members. I look forward to future opportunities to promote Florida's world-class cattle industry.

Sincerely,

Adam H. Putnam
Commissioner of Agriculture

Fresh From Florida. 1-800-HELPFLA (850) 488-3022 www.FreshFromFlorida.com

FLORIDA CATTLEMEN'S FOUNDATION, INC.
PO BOX 421929
KISSIMMEE, FL 34742-1929
(407) 846-6221

November 27, 2012

Greetings:

Florida Cattle Ranching: Five Centuries of Tradition is the first comprehensive exhibition to present the history and culture of Florida's rich legacy of cattle ranching. It serves to accomplish two of the primary missions of the Florida Cattlemen's Foundation: preserving our state's ranching history, and educating the public about the Florida beef industry.

This publication and the exhibit it presents are the products of a team effort by hundreds of folks—more than I could ever thank individually in this letter. Thank you all for your generous support.

The Foundation is especially grateful to its board of directors, who give their valuable time and talents so freely; Commissioner Adam Putnam, for leading the Department of Agriculture and Consumer Services in finding a long-term home for the exhibit at the Florida State Fair; the board of directors of the Florida State Fair Authority, whose vision and enthusiastic support made the new exhibit venue possible, particularly chairman Doyle Carlton III, who continues the Carlton family tradition of supporting the State Fair so generously over three generations; and all those who contributed funds to this most worthwhile project.

Join us as we celebrate the rich history and vibrant culture of the oldest cattle ranching state in America!

Sincerely,

Jim Strickland

Jim Strickland
Chairman

Preserve ★ *the Heritage of the Florida Cattle Producers and All Property Owners*

Promote ★ *Environmental Stewardship* ★ *a Sustainable Cattle Industry*
through scholarship, research, educational programs, and leadership development

Introduction

Florida's cattle industry, one of oldest and largest in the nation, is vital to the state's well-being. Ranching is an essential economic activity that preserves many aspects of the natural landscape, protects water resources, and maintains areas used by wildlife or for recreation. Yet few know about Florida's unique ranching traditions, which have been adapted to the subtropical climate and influenced by the state's distinctive history.

In Florida, those who own or work cattle traditionally have been called cowmen. In the late 1800s they were often called cow hunters, a reference to hunting for cattle scattered over the wooded rangelands during roundups. At times the terms cowman and Cracker have been used interchangeably because of similarities in their folk culture. Today the western term "cowboy" is often used for those who work cattle.

Cow skulls on fence. Okeechobee, 2008. Photo by Robert L. Stone, courtesy of the Florida Folklife Program/Florida Department of State.

A collection of cow skulls adorns a fence at the Dixie Ranch.

Brahman, Brahman-Angus crossbreed, and egrets. Vero Beach, 2008. Photo by Bob Montanaro.

Cattle egrets among Brahman (light) and Brahman-Angus crossbred (dark) cattle at the Treasure Hammock Ranch. A common sight in Florida, the graceful cattle egrets came from Africa via South America. They eat insects that cattle flush from the grass as they graze.

Herd ng cattle through water, Horse Creek Ranch, Hardee County, 2012.
Photo by Carlton Ward, Jr.

Colonial Florida

Florida ranching has evolved from many different cultural traditions, though the most important sources were the marshy coastal areas of Andalusia, Spain, and the hill regions of Britain and Ireland. In Andalusia, ranchers living in towns hired cow hands (vaqueros), who marked or branded the cattle, managed them from horses, and moved them to different locations during the year. They later brought long-horned Andalusian cattle to the Americas. In highland Britain and Ireland, herders marked or branded cattle for identification, penned them at night for protection, and moved them to different pastures during the year. In the fall, the animals were sold to drovers, who used dogs and whips to drive them to markets or slaughterhouses. The Spanish and British took these traditions to the West Indies, where they were adapted to the tropical climate and combined to create ranching systems used throughout the Americas.

Florida's Andalusian/Caribbean cattle were the first in today's United States. Some scholars believe that cattle brought by the expeditions of Ponce de Leon in 1521 and Don Diego de Maldonado in 1540 escaped and survived in the wild. Organized ranching began with the founding of St. Augustine in 1565, when cattle from Spain and Cuba formed the basis of herds that fed the garrison and surrounding communities. In addition to herds owned by the Spanish and Indians, wild cattle flourished in the rangelands and prairies. Eventually Spanish colonists began exporting cattle to Cuba. During the 1600s, Spanish clergy raised cattle at the missions, where many Native Americans learned to tend them.

By 1700 Florida contained approximately 34 ranches and 20,000 head of cattle. After British-Creek Indian raids in 1702 and 1704 devastated Florida cattle ranchers, Indians sustained cattle raising

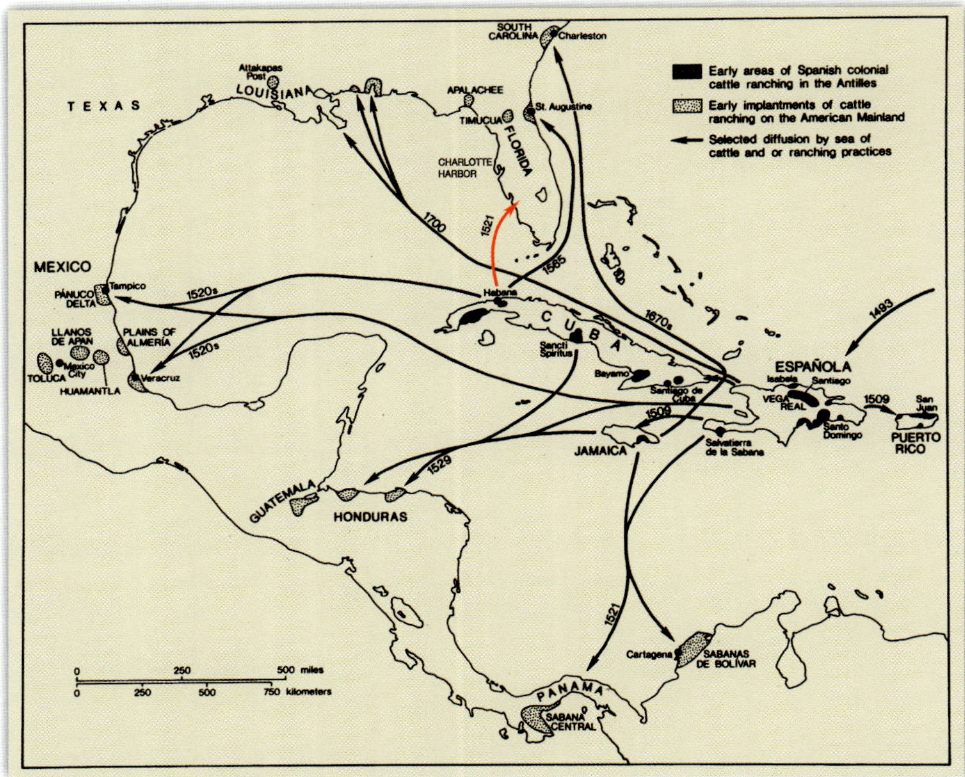

Colonial Cattle Ranching in the West Indies and Diffusion to the Mainland.

Based on the map on page 66 of *North American Cattle-Ranching Frontiers: Origins, Diffusion, and Differentiation* by Terry Jordan (Albuquerque: University of New Mexico Press, 1993), which has been revised to include the cattle brought by Ponce de Leon to southwest Florida in 1521.

Old Spanish Brands. Drawn by Joe A. Akerman, Jr.
Courtesy of the State Archives of Florida, image rc5555.

Florida Cracker Cow. Sponsored by the Florida Cattlemen's Foundation and the Florida Department of Agriculture and Consumer Services. Photo by Ray Stanyard.

Florida Cracker Cattle descend from criollo *cattle brought to Florida in the 16th century by the Spanish. Settlers of British descent who were known as Crackers moved into Florida— primarily from Georgia and South Carolina—in the late 18th and early 19th centuries. The independent, hardy, and self-reliant Crackers rounded up wild Spanish cattle to form their own herds. These cattle, which became known as Cracker Cattle, were the foundation of Florida's immense cattle industry. Cracker Cattle were able to withstand the heat and insects of the Florida environment but were relatively small and yielded less beef compared to "modern" breeds. In the late 19th and 20th centuries, Florida cattlemen sought to improve their herds by introducing breeds such as Red Devon, Hereford and Brahman cattle. The Cracker Cattle were being slowly bred out of existence. Through the efforts of the Florida Department of Agriculture and Consumer Services the Florida Cracker Cattle Association was formed in 1988. Today, the future of Cracker Cattle as a heritage breed is ensured through herds maintained by the Department and private owners.*

The small, lean body, head shape, and horns distinguish Florida Cracker cattle from other breeds.

Spanish stirrup. Brass, ca.16th–17th century. Loaned by the Florida Bureau of Archaeological Research. Photo by Ray Stanyard.

This is the only verifiable colonial Spanish stirrup in Florida.

in Florida. Many had established large herds of wild cattle and stock acquired from the Spanish. Cattle herding was vital culturally and economically to the early Seminoles, as attested to by the name of their leader, Cowkeeper (ca. 1710-1783). They remained Florida's major livestock producers throughout most of the 1700s.

During British rule (1763-1783), English planters and Creek Indians in west Florida owned substantial herds. Cowmen from Georgia and the Carolinas spread into north Florida during that period.

In early Florida, Europeans, Americans, and Indians stole cattle from each other. Rustling became particularly widespread by the second half of the 18th century, and was one of the elements that led to the Seminole Wars.

During the periods of Spanish colonization (1565-1763, 1783-1819), the main ranching areas were Payne's Prairie in Alachua County, Tallahassee-St. Marks, inland from St. Augustine along the St. John's River, and, later, Pensacola. Hides, tallow, and sun-dried beef were processed for the St. Augustine garrison and for export. Ranchers owned land, cattle, and horses; cowhands came from more modest economic backgrounds and many were Native American.

Handmade wrought iron British-style riding stirrup, ca. late 18th–early 19th century. Loaned by the Florida Bureau of Archaeolog ical Research. Photo by Ray Stanyard.

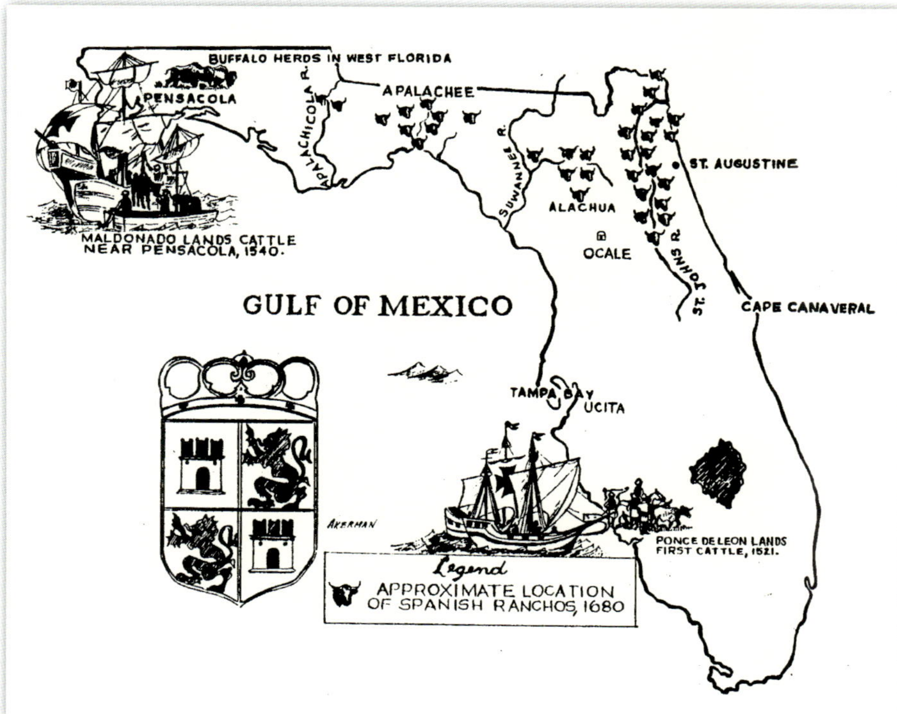

Location of Spanish ranchos in Florida ca. 1680. From *Florida Cowman: A History of Florida Cattle Raising* by Joe A. Akerman, Jr. Kissimmee: Florida Cattlemen's Association, 1976. Page 4.

Spanish riding spur. Reproduction of a spur found on the Atocha. Loaned by the Florida Bureau of Archaeological Research. Photo by Ray Stanyard.

The Nuestra Senora de Atocha and the Santa Margarita were Spanish galleons that sank near Key West during a hurricane on September 6, 1622.

Riding spur. Iron. Pre–1704. Loaned by the Florida Bureau of Archaeological Research. Photo by Ray Stanyard.

This light, delicate spur was found at the San Juan de Espalaga mission site, Jefferson County.

Horseshoe. Wrought iron. Loaned by the Florida Bureau of Archaeological Research. Photo by Ray Stanyard.

This is the only horseshoe from a Spanish ship wreck ever recorded. It was found on the San Miguel de Archangel, which sunk off the coast of Jupiter in 1659.

Snaffle bit. Tin-coated iron, ca. 1760. Loaned by the Florida Bureau of Archaeological Research. Photo by Ray Stanyard.

Probably a trade item, this bit was found at the Flintlock Site, a Seminole site on the Apalachicola River, Jackson County. Other trade items found at this site include alloyed silver trinkets and muskets.

Steamship at Wharf. Punta Rassa, 1890s. Courtesy of the State Archives of Florida, image rc19392.

At the wharf, cattle were crowded into every available spot on board schooners or steamships that plowed the waters between Punta Rassa, Tampa, St. Andrews Bay, Charlotte Harbor and Cuba.

The 19th Century: Struggle and Sacrifice

When the U.S. took possession of Florida in 1821, it was described as a "vast, untamed wilderness, plentifully stocked with wild cattle." Florida "scrub" or Cracker cattle were descended from the mix of Spanish and British breeds. These hardy creatures survived on native forage, tolerated severe heat, insect pests, and acquired immunity to many diseases.

Early Florida cowmen survived in difficult conditions. They fought off panthers, wolves, bears, and cattle rustlers. Cowmen spent weeks or months on cattle drives across difficult marshes and dense scrub woods, often enduring burning heat, torrential thunderstorms, and hurricane winds. From central Florida they sometimes drove cattle as far as Jacksonville, Savannah, and Charleston. This gradually changed in the 1830s when the cattlemen re-established trade with Cuba, and Tampa, Punta Gorda, and Punta Rassa became important export ports.

The number of cattle increased rapidly from the 1840s until the Civil War. Florida was second only to Texas in per capita value of livestock in the South. After the Armed Occupation Act of 1842, cattlemen from the overstocked states of Georgia, Alabama and the Carolinas homesteaded 200,000 acres in Florida. Some seized

Cattle Drive at Bartow. 1890s. Courtesy of the State Archives of Florida, image rc02673.

This early photograph was identified as follows: "At the far left is Crayton Parker, in the middle is Tom Smith on a horse named Boomerang and at the right is Aunt Jeanie feeding hay to a cow."

Scrub Cattle at Punta Rassa, early 1900s.
Courtesy of the Southwest Florida Museum.

Jacob Summerlin. Courtesy of the State Archives of Florida, image rc02467.

Jacob Summerlin *(1820-1893)*

Reputedly the first child born in the Florida Territory, Jake Summerlin was said to have started working cattle and cracking whips by seven years of age. At 16, he travelled south to central Florida, where he earned his fortune raising cattle in the Kissimmee and Peace River areas. Summerlin became one of the state's wealthiest men before he reached 40. He and his partners sold cattle to Cuba and the U.S. Naval Base in Key West. He purchased large land parcels in southwest Florida, including a wharf at Punta Rassa. During the Civil War, Summerlin smuggled beef to the Confederates by shipping them out of present day Charlotte Harbor, then later sold cattle to Union soldiers at Ft. Myers. After the war he donated land to establish a school in Bartow. In Orlando, he opened the Summerlin Hotel, donated the land for Lake Eola Park, and became the City Council's first president.

range territory that the Seminoles had been forced to relinquish as a consequence of the Seminole Wars. The newcomers often brought foundation herds that interbred with wild scrub cattle. Few cattlemen owned grazing land since there was extensive open range. By mid-century, ranchers were running large herds on the extensive open range in central and south Florida.

Wars provided an economic boost for Florida cattlemen, who provisioned armies during the Seminole, Civil, and Spanish American Wars. Although the Civil War disrupted the Cuban trade, Florida cowmen became beef suppliers to both armies. Hides, tallow, and meat from Florida were so important for the Confederacy that a Cow Cavalry was organized to protect herds from Union raiders.

Fighting over a Stolen Herd. From Frederic Remington (1861-1909), "Cracker Cowboys of Florida," Harper's New Monthly Magazine, v.91:543, August 1895. Courtesy of the State Archives of Florida, image rc02054.

The best-known artist and chronicler of the western cowboy, Frederic Remington, visited Florida in 1895. In early Florida, Europeans, Americans, and Indians stole cattle from each other. Rustling has continued to be an irritant throughout Florida into the present—with some of the worst occurring after World War II.

A Cracker Cowboy. From Frederic Remington (1861-1909), "Cracker Cowboys of Florida," *Harper's New Monthly Magazine*, v.91:543, August 1895. Courtesy of the State Archives of Florida, image rc02055.

Bone Mizell, on his mount and his dog at his side, was sketched by Remington when he came to Arcadia in 1895 to illustrate and write about the Florida frontier.

Bone Mizell

Morgan Bonaparte Mizell, known to his fellow cowmen and the law as just Bone, was a true legend in his own day, held in high esteem by his cowhunter peers for his free-spirited, hard-drinking, fun-loving approach to living. He was said to "outrope, outride, outshoot and outdrink" any cowman in Florida. It is thought that the folk hero offered his unpredictable antics to all as some distraction from their hard lives as range riders.

While tales about Bone Mizell have been exaggerated over the years, his real life was colorful and represents well the eccentric lifestyle of a Cracker cowboy. He rustled cattle, marked his livestock with his teeth (by notching the cow's ears), and rode horses into bars and ordered drinks from his saddle. Once, he hitched a big-top tent to a passing freight train when a traveling circus banned him for attracting too much attention. Known for his wild spending sprees and spur-of-the-moment generosity, Mizell was hardly driven by money and did not eagerly aspire to be wealthy.

Bone Mizell's hard-living lifestyle took its toll and on July 14, 1921 at the age of 58 he died drunk in a railroad depot at Fort Ogden, Florida. Mizell and other legendary Cracker cowhunters will forever be etched in Florida folklore as enduring symbols of the way things used to be on the Florida range.

Dana Ste. Claire, *Cracker: The Cracker Culture in Florida History*, Daytona Beach: Museum of Arts & Sciences, 1998, pp. 181-183.

"Them that's got, has to lose. Them that hasn't, kaint."

—Bone Mizell, ca. 1900.

After the Civil War there were still wild cattle in Florida, as well as attractive markets to the north and south. During the next three decades, trade boomed with Cuba, Key West, and Nassau, and Florida became the nation's leading cattle exporter. From 1868 to 1878, ranchers received millions of dollars in gold doubloons for more than 1.6 million cattle exported to Cuba. The Cuban commerce provided income to cattlemen, merchants, and shippers, and contributed to the state's recovery from Reconstruction-era depression.

A bit and a pair of spurs, ca. late 19th century. Loaned by Billy Davis. Photos by Ray Stanyard.

These inexpensive, utilitarian items are typical of those used by Cracker cowboys of the period, possibly originally purchased from Montgomery Ward or Sears Roebuck and Co. catalogues.

Bit, Star brand, Buermann's Dog-leg. Cast iron.

Spurs, Buermann's Eureka Star brand. Sheet metal.

The cowboy on the opposite page is using a bit and spurs identical to these.

Cowboy with Cracker horse at an open range roundup near Fort McCoy, ca. 1910. Courtesy of the State Archives of Florida, image n045020.

A Bit of Cow Country. From Frederic Remington (1861-1909), "Cracker Cowboys of Florida," Harper's New Monthly Magazine, v.91:543, August 1895. Courtesy of the State Archives of Florida, image rc02057.

The drawing depicts a 19th century cow camp.

Two bits and a pair of spurs, ca. late 19th century. Loaned by Billy Davis. Photos by Ray Stanyard.

Bit, McChesney Gal-leg. Steel.

Bit, North & Judd Anchor brand. Cast iron.

Spurs, North & Judd Anchor brand. Steel.

McClellan Saddle with saddle pockets. Leather with brass fittings, ca. 1904. Loaned by James Levy. Photo by Ray Stanyard.

Thousands of military saddles like this McClellan were left over at the end of the Civil War, Spanish-American War, and World War I. Many were utilized by American stockmen, particularly in Florida, as the split seat afforded good ventilation. Several different McClellan styles were manufactured, but only the model 1913 "mule packer's saddle" was made with a saddle horn. When roping became more common in Florida, early cowmen would sometimes attach a small horn to the front of the pommel. As roping became more common, McClellan saddles faded from popular use. Today working cowboys rely almost entirely on western saddles.

Barbed Wire. Steel. Loaned by the Florida Cattlemen's Association (FCA). Donated to the FCA by Linda A. Dorough. Photo by Ray Stanyard.

Florida cattlemen used barbed wire long before fencing was required by state law in 1949. Fencing became necessary during the 1920s and early 1930s during the tick eradication program, which required dipping all cattle every 14 days. Fencing was also used to control breeding, which improved herds.

From top:

638B Kelly Thorny Common, pat. February 11, 1868
651B Stover Regular Corsicana Clip, pat. June 29, 1875
39B Glidden Twisted Oval, pat. August 22, 1876
175B Wing Two Staple, pat. February 26, 1878
162B Reynolds Necktie, pat. May 14, 1878
693B Scutt Arrow Plate, pat. June 18, 1878
722B Crandal Zigzag, pat. November 4, 1879
992B Allis Medium Width Buckthorn, pat. July 26, 1881
306B Ellwood Parallel Strands& Reverse Wrap Barb, pat. January 31, 1882
774B Stubbe Large Plate, pat. October 23, 1883
861B Cady Barbed Link, One & One-Half Wrap, pat. January 22, 1884
9B Briggs Slanted Kinked, pat. July 1, 1884
733B Hodge Spur Rowel on Twisted Strands, pat. August 2, 1887

Mitchell L. Kolbe (1955–). Get Behind. Bronze sculpture. Commissioned by the Florida Heritage Historical Society. Loaned by Connerton, a new town in Pasco County. Photo by Ray Stanyard.

This historically accurate work portrays a cowman returning to camp at the end of a long day, calling his catch dog off a subdued cow that wandered from the herd with the command, "Get behind." Kolbe worked with Bronzart Foundry, Inc. in Sarasota, to cast an edition of 27 sculptures from a wax model. For more information about Get Behind, contact Mitch Kolbe at mitchkolbe@yahoo.com or the Florida Heritage Historical Society, dconner@connergroup.net.

76

20th Century: Ranching Modernizes

The 20th century brought changes affecting cattle ranching infrastructure and occupational traditions. As Cuban demand declined early in the century, ranchers turned to domestic markets and cattle were shipped by rail throughout the country.

Modern technology brought many positive changes. As early as 1858, ranchers attempted to improve the quality of their herds with Brahman cattle, originally from India. Since the 1930s, ranchers have successfully cross-bred native cattle with Brahman, Angus, Hereford, Shorthorn, Charolais, and Limousin to improve size, resistance to heat and insects, hardiness, and meat quality. This led to the development of hybrids such as the Braford and Brangus breeds.

Ranchers struggled with and overcame severe insect pests, such as Texas ticks and screw-worms, through medical and scientific research. In the early 1900s, Texas tick fever entered Florida with cattle from other states. By the 1920s, the State Livestock Board made it mandatory for ranchers to construct vats and dip cattle in an arsenic solution every two weeks. Although expensive and

Tick Inspection Station at the Baker County Line. Early 1900s. Courtesy of the State Archives of Florida, image pr01403.

Brahman bull. Bradenton, 1900s. Courtesy of the State Archives of Florida, image pc5941.

Brahman cattle are well-suited to Florida because they are heat tolerant, mature early, and are larger and more disease-resistant than many breeds. This early Brahman bull, Emperor Jr. 10th, owned by T.P. Chaires Jr. of Bradenton was said to be one of the outstanding Brahman Bulls of the time.

Brahman cattle, Hardee County, April 2008. Photo by Carlton Ward, Jr.

Brahman cattle head for the cow pens on the Bar Crescent S Ranch. The fifth generation of the Smith family continues the ranching tradition established by their great-great grandfather, Doyle Carlton, governor of Florida from 1929-33.

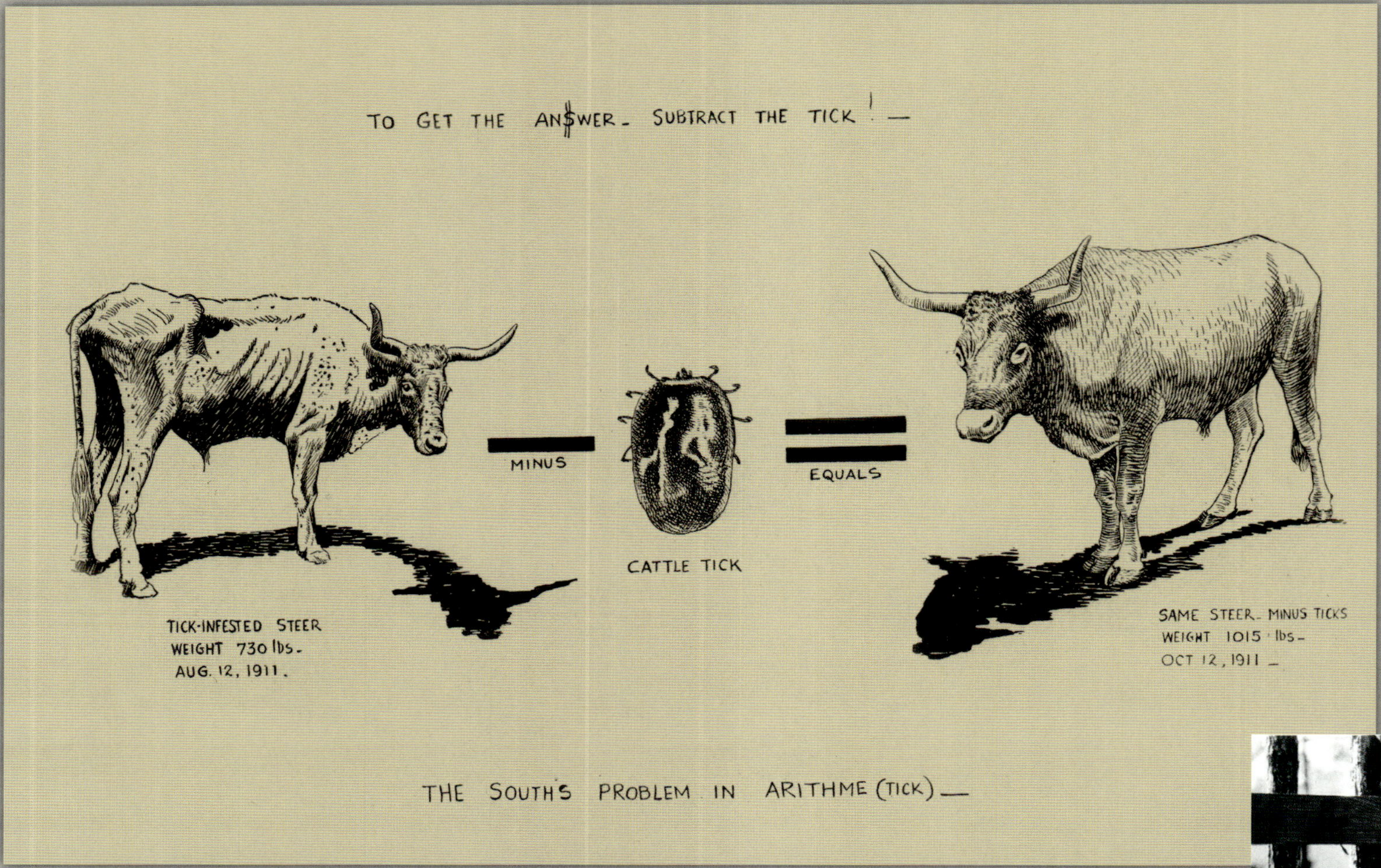

Effects of Ticks on Cattle. 1913. Courtesy of the State Archives of Florida, image pr01404.

Duval County man dipping and paint marking cattle between 1923 and 1929. Courtesy of the State Archives of Florida, image pr01396.

Emasculator, Burdizzo brand. Iron with wood handles, ca. 1950. Loaned by Sean Sexton. Photo by Ray Stanyard.

This implement was used during the screw worm epidemic to render male calves sterile without making an incision, which would attract screw worm flies. Unfortunately, it was not 100% effective. Castration with a sharp knife is the common practice today.

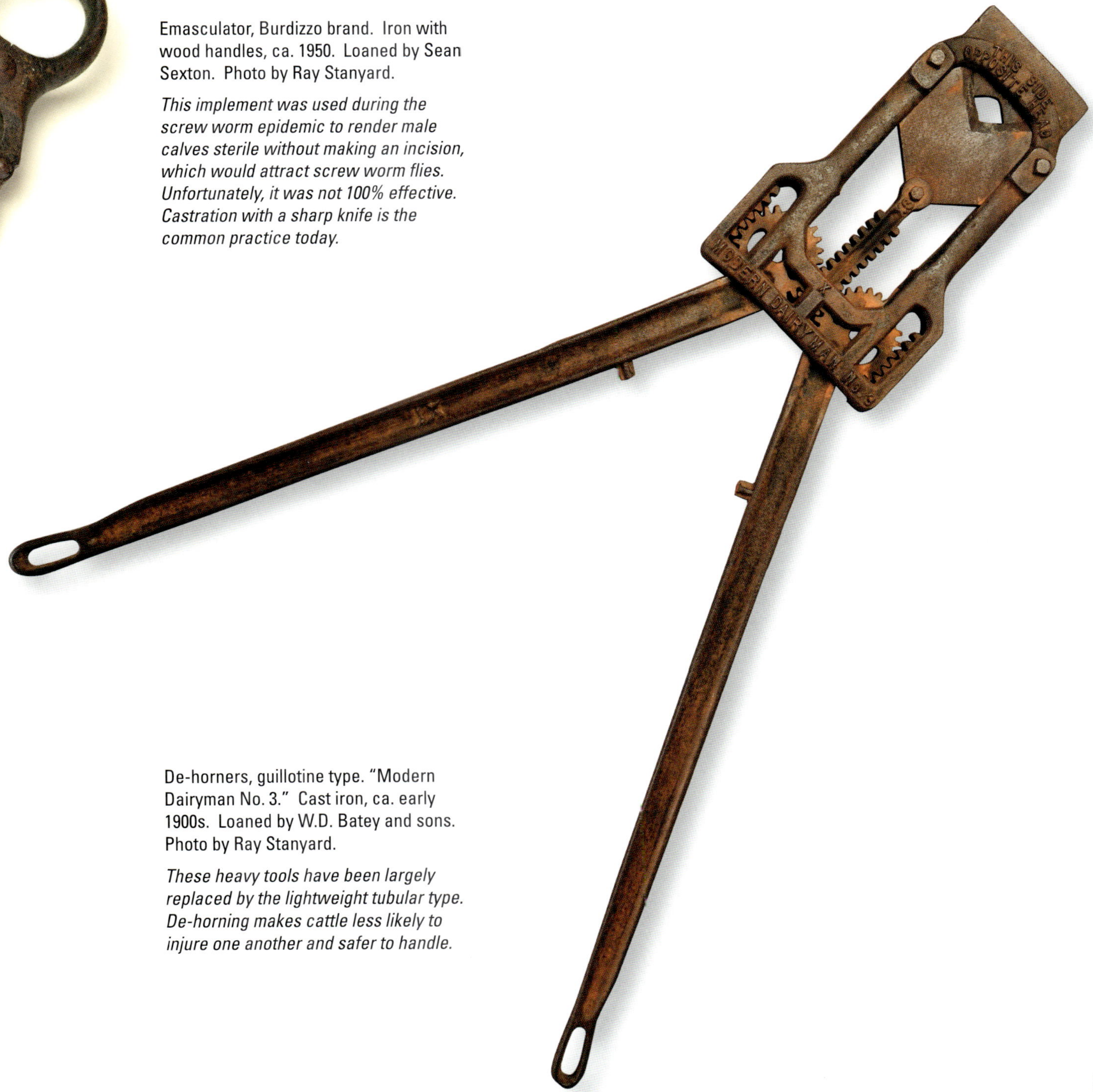

De-horners, guillotine type. "Modern Dairyman No. 3." Cast iron, ca. early 1900s. Loaned by W.D. Batey and sons. Photo by Ray Stanyard.

These heavy tools have been largely replaced by the lightweight tubular type. De-horning makes cattle less likely to injure one another and safer to handle.

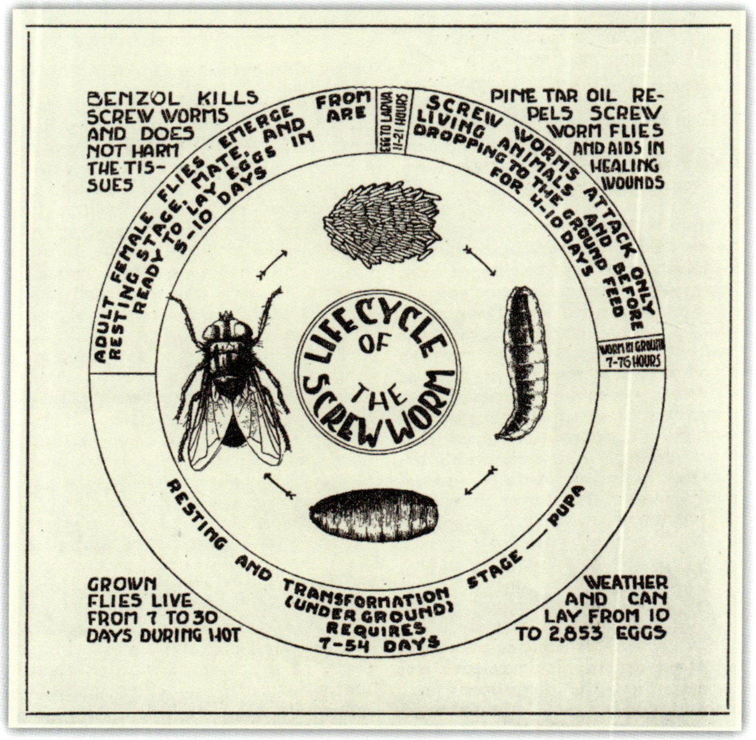

Screwworm cycle, *The Florida Cattleman and Livestock Journal.* October 1949.

difficult, the effort curbed the problem. Fencing became widespread in order to separate treated from untreated animals. The fever resurfaced in 1935 and 1960, resulting in intense eradication programs implemented by Florida's Department of Agriculture.

In 1949 the Florida Legislature passed the Warren Act, requiring ranchers to fence their pastures. The fence law struck a blow against small ranchers who relied upon grazing cattle on open tracts of land that they could not afford to buy.

In the mid-20th century, many ranchers improved nutrition by upgrading their pasture grasses. This made it possible to raise a cow on three or four acres, in contrast to the 40 acres previously required. To further maximize growth, ranchers provided their cattle with supplementary feed at crucial points in the yearly production cycle. Dried citrus pulp pellets, produced from byproducts at citrus processing plants, have been used as feed since the 1920s.

Barry's Screw Worm Killer. Liquid in steel can, ca.1940s. Loaned by Julianne Barry. Photo by Ray Stanyard.

Barry's Screw Worm Killer, made in Newberry, was one of two popular screw worm medications marketed statewide by north-central Florida companies. The other was Smearex, which was made in Ocala. During the screw worm epidemic every animal had to be frequently checked for cuts and scratches and all wounds treated with medication.

Fighting screwworms. Sebring, 1958. Photo by Jim Stokes. Courtesy of the State Archives of Florida, image c028500.

A pilot prepares boxes of sterile screwworm flies to be dropped on a cattle ranch. Within a short time this program eradicated one of the most serious pests in the history of cattle ranching. It is estimated that the screwworm damage to livestock in the Southeast was about $20 million annually from 1935-1959.

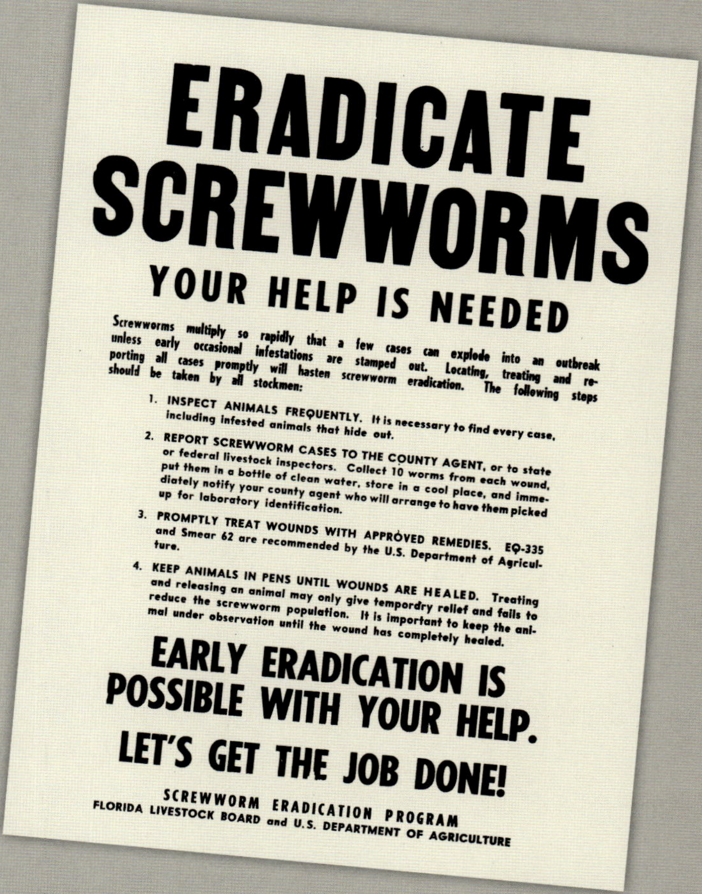

ERADICATE SCREWWORMS

YOUR HELP IS NEEDED

Screwworms multiply so rapidly that a few cases can explode into an outbreak
unless early occasional infestations are stamped out. Locating, treating and re-
porting all cases promptly will hasten screwworm eradication. The following steps
should be taken by all stockmen:

1. INSPECT ANIMALS FREQUENTLY. It is necessary to find every case,
including infested animals that hide out.
2. REPORT SCREWWORM CASES TO THE COUNTY AGENT, or to state
or federal livestock inspectors. Collect 10 worms from each wound,
put them in a bottle of clean water, store in a cool place, and imme-
diately notify your county agent who will arrange to have them picked
up for laboratory identification.
3. PROMPTLY TREAT WOUNDS WITH APPROVED REMEDIES. EQ-335
and Smear 62 are recommended by the U.S. Department of Agricul-
ture.
4. KEEP ANIMALS IN PENS UNTIL WOUNDS ARE HEALED. Treating
and releasing an animal may only give temporary relief and fails to
reduce the screwworm population. It is important to keep the ani-
mal under observation until the wound has completely healed.

EARLY ERADICATION IS POSSIBLE WITH YOUR HELP.

LET'S GET THE JOB DONE!

SCREWWORM ERADICATION PROGRAM
FLORIDA LIVESTOCK BOARD and U.S. DEPARTMENT OF AGRICULTURE

Screwworm eradication flyer by Florida Livestock Board and U.S. Department of Agriculture. 1940s. Courtesy of Florida Cattlemen's Association.

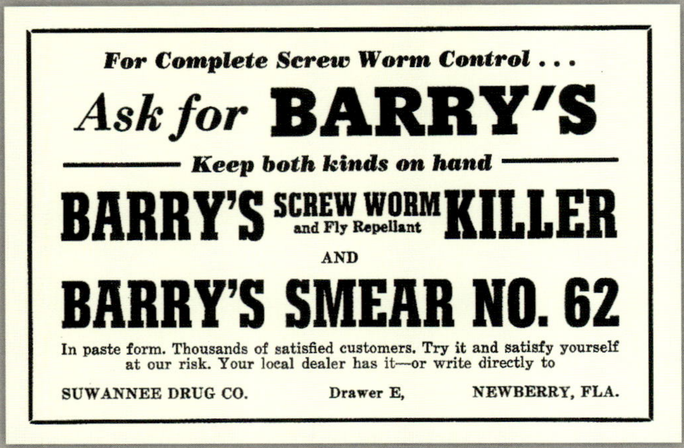

Barry's Screwworm Killer advertisement. From *The Florida Cattleman and Livestock Journal.* February, 1945.

Cookware and utensils used by cow hunters in the late 19th and early 20th centuries. Loaned by W. D. Batey and sons. Photos by Ray Stanyard.

Skillet, spider type. Cast iron, ca. early 1900s.

The legs keep the skillet above the wood fire.

Forks, dinner. Steel with wood handles, ca. 1860s.

Probably from Civil War mess kits.

Plate, dinner. Porcelain over tin, ca. 1860s.

Spoon, wooden.

Marks and Brands Record, Polk County, vol. 2 (1938–1945). Loaned by Polk County Historical Museum. Photo by Ray Stanyard.

Before registration became statewide in 1945, livestock brands and earmarks were recorded by each county. Cattle ears are represented by a figure-eight diagram. The brands and marks of many pioneer cattlemen recorded in this book reflect Polk County's rich history of cattle ranching.

Early barbed wire fencing in Osceola County, Kissimmee. 1915. Courtesy of the State Archives of Florida, image rc11287.

EAR MARK	OTHER MARKS
	Roberts, Dr. T. H. Records his Mark as underslope in right ear and two underbits in left ear. Branded with flying Bob-White This August 26-1939. D. H. Sloan Jr. Clerk
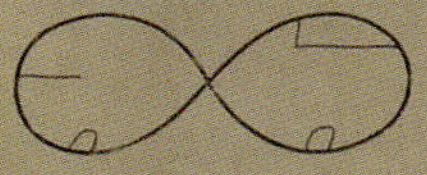 V C.	V. C. Ryals records his mark as split under bit in one ear and upper square under bit in other ear. Brand V C. This March 14-1940. D. H. Sloan Jr. Clerk
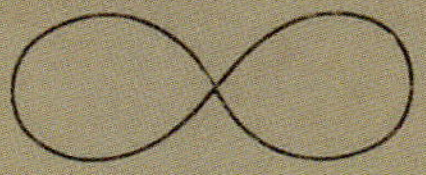 Ⓡ	Rhoads, C. S. Records his stock brand as follows: Ⓡ. This Sept. 23-1940. D. H. Sloan Jr. Clerk
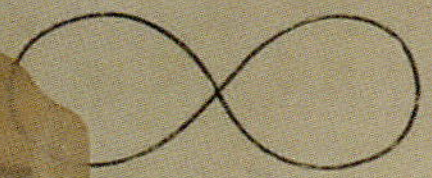	
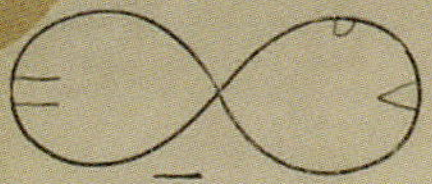 Ā	Rhoden, Arnold, records his mark and brand as follows: Swallowfork upper bit in one ear, two straight splits in the other ear, Branded Ā. This Jan. 12, 1942. D. H. Sloan Jr. Clerk
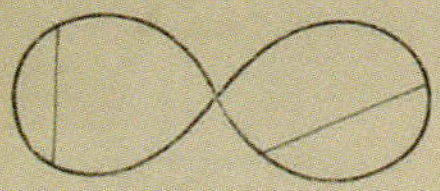 VI	Roberts, Dr. T. H., Records his mark as crop in left ear and underslope in right ear. Branded with Roman VI. This October 8, 1942. D. H. Sloan, Jr., Clerk

EAR MARK	OTHER MARKS
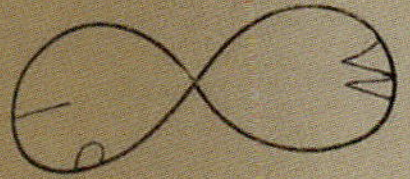	Rawlerson, J. N., records his mark as two swallow forks in one ear and straight split under bit in the other ear. This Jan. 25-1943. D. H. Sloan Jr. Clerk
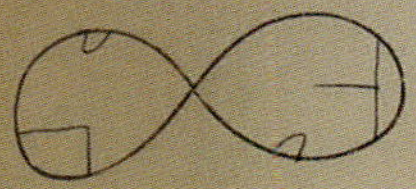	Rawlerson, J. N. records crop split under bit in one ear, and under square upper bit in other ear. This Jan 25-1943. D. H. Sloan Jr. Clerk
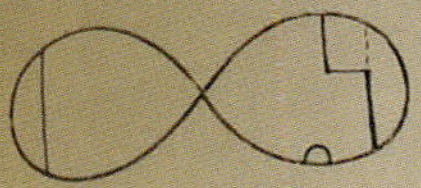	Revis, Rosa B., of Davenport, records her hog and cattle mark as: Crop in left ear and crop, upper square and underbit in right ear. This Sept. 17, 1943. D. H. Sloan, Jr., Clerk
 R̄	Rhoden, Hardie, records his mark and brand as follows, Staplefork in one ear, and swallow fork under half Fleur de lis in the other ear Brand R̄. This Feb. 21-1944 D. H. Sloan Jr. Clerk
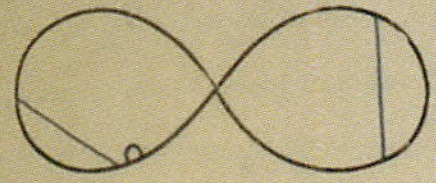 WJ	Rhymes, W. J. of Homeland, records his mark and brand as follows: Underslope, underbit in one ear and crop in the other Branded WJ. This May 5, 1944 D. H. Sloan, Jr., Clerk
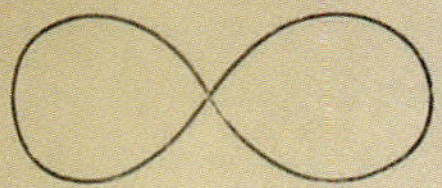	

CORRAL
BRAND
PINELLAS COUNTY FRUIT
CLEARWATER GROWERS ASSN.
CLEARWATER • FLORIDA
©1936 C.G.A.

ROPER BROS.
WINTER GARDEN,
FLORIDA
ROPER
BRAND

REG. U.S PAT. OFF.
Lasso
BRAND
TRUBLU
WINTER HAVEN
FRUIT SALES CORP.
WINTER HAVEN,
FLORIDA

RANGER
BRAND
VAUGHN-GRIFFIN PACKING CO.
HOWEY, FLORIDA
VAUGHN-GRIFFIN PACKING CO.
HOWEY, FLORIDA

Citrus labels. Courtesy of the State Archives of Florida.

Cattle ranching and cowboy themes have been incorporated into many citrus labels, a practice that demonstrates the close connection of the citrus and cattle industries in Florida.

Cattle feed produced from Minute Maid Company's citrus pulp. 1950s. Courtesy of the State Archives of Florida, image rc15569.

Cattle feed made from pulp, membranes, seed, and rind discharged during the manufacture of concentrate converted a disposal problem into a profitable operation.

Waverly Brands advertisement. *The Florida Cattleman and Livestock Journal.* January, 1949.

Feeding cattle whole citrus is a practice continued by some cattlemen today.

Arsenic test kit. Cooper's "Cattle Dip" Tester, William Cooper & Nephews, Chicago, ca. 1920s. Wood, glass, cotton, and other materials. Loaned by David Hunt. Photo by Ray Stanyard.

This test kit from the Florence Akins Ranch in Polk County was used to measure the strength of the arsenic solution used for dipping cattle during the tick eradication program of the 1920s and 1930s.

Cow killed by an automobile. Volusia County, 1920s. Florida State Archives. Courtesy of the State Archives of Florida, image rc11290.

Before fencing became mandatory in 1949, automobiles and cows sometimes collided. The result was often disastrous.

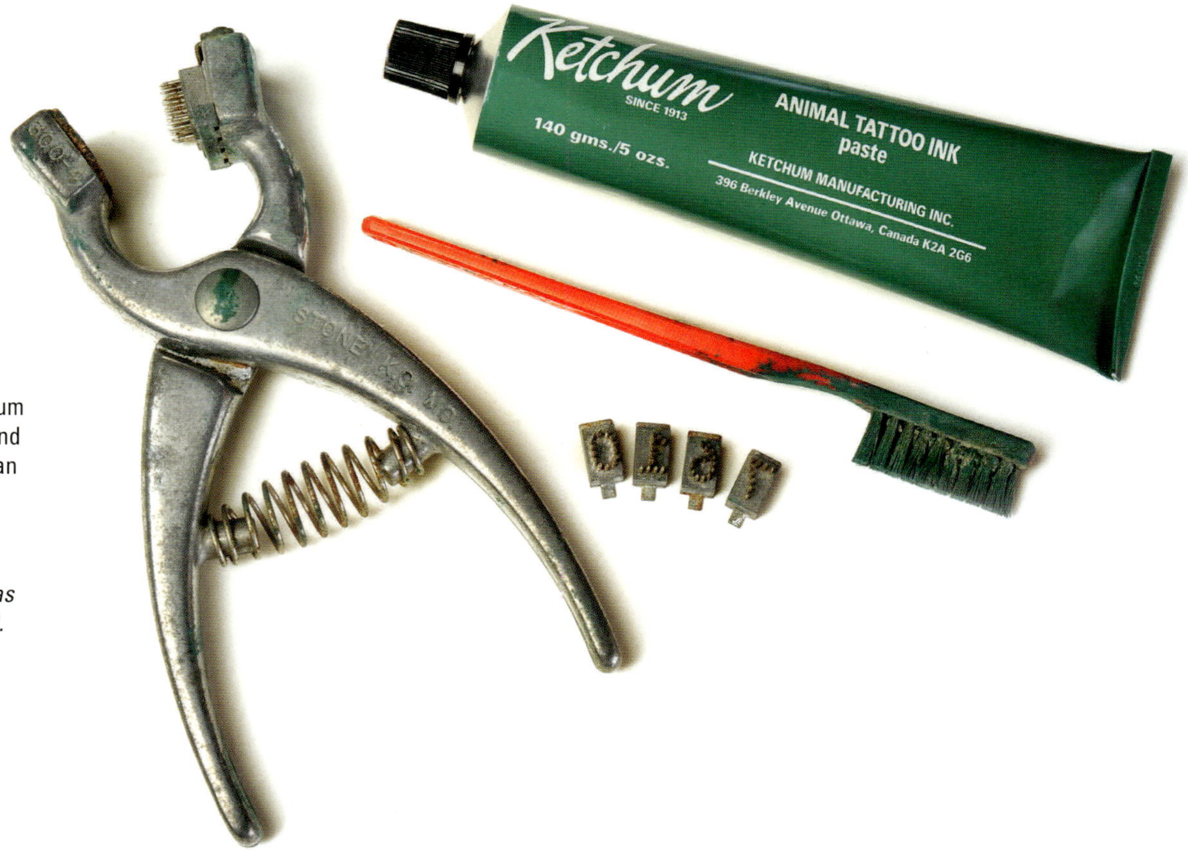

Cattle tattooing kit: cast aluminum pliers; Ketchum brand "Animal Tattoo Ink;" needle-type letters and numerals; and plastic toothbrush. Loaned by Sean Sexton. Photo by Ray Stanyard.

The ears of cattle are tattooed to track tests for brucellosis infection, also known as contagious abortion and Bang's disease. In 2001, Florida was granted brucellosis class-free status by the U. S. Department of Agriculture.

Seminole Cattle Ranching

The Seminoles' relationship with cattle has endured for centuries. Prior to their arrival in Florida and subsequent designation as Seminoles, Creek Indians in Alabama and Georgia were engaged in cattle production. By 1740 the followers of the Oconee Creek leader, Cowkeeper, established Florida's largest settlement in the Alachua area. Heavily involved in a cattle economy, by 1775 the Seminoles were working cattle herds alleged to number thousands of head on Paynes Prairie using trained cow-dogs.

Slavery was a long-term cultural practice among the southeastern Indians. Black slaves fleeing to Spanish-held Florida often lived with the Seminoles, affording them a major source of manpower for such pursuits as cattle production. There was constant trouble between the Seminoles and Georgians over the ownership of cattle and runaway slaves on the Florida/Georgia border. These conflicts sparked U.S. aggressions against the Seminoles in the First Seminole War (1818).

Continued aggressions between the Seminoles and European-American settlers over cattle and grazing lands contributed significantly to starting the Second (1835-1842) and Third Seminole Wars (1855-1858), which reduced the Seminole population to around 200.

After the Civil War, Seminole families in Big Cypress and around Lake Okeechobee raised cattle despite harassment by rustlers. In the early 20th century, cattle posed such a liability that few Seminoles raised them. Without strife over rangelands, the Seminoles and Cracker cattlemen got along well. Ranchers sometimes hired Seminoles, whom they considered to be excellent cattlemen.

A new era of Seminole cattle ranching began in the 1930s, when the Dania and Brighton Seminoles acquired starter herds. The Seminole Tribe established the Indian Livestock Association in 1939. In 1944, they created separate cattle enterprises for Brighton and Big Cypress, with the Central Tribal Cattle Organization providing general supervision. Seminoles banded with other Native American stockmen in 1974 to form the National American Indian Cattlemen's Association.

Today, the Seminole Tribe is one of Florida's leading beef producers. Sophisticated video auctions promote tribal cattle across the nation. The success of the cattle program has allowed the Tribe to pursue other successful economic ventures.

Le Anne Billie with horse. Big Cypress Seminole Indian Reservation. Photo by Carlton Ward, Jr.

Lariat, or Lasso. Nylon. Immokalee, 1997. Loaned by Stanlo Johns. Photo by Ray Stanyard.

This coiled lariat is made of three-strand nylon hard twisted rope. The Seminole Brand rope factory began manufacturing in 1996 and closed about two years later.

Bumper sticker advertisement for Seminole Ropes. Paper. Circa 1997. Donated by Stanlo Johns. Photo by Robert L Stone.

Cattle branding irons, ca. 1940–1960. Steel. Loaned by Stanlo Johns. Photo by Ray Stanyard.

Brands shown here are "ID," "J," and "HG." ID stands for "Indian Department." It was used from the early 1940s until around 1953, when the cattle were distributed to individual herd owners. Toby Johns (Panther clan) used the J brand, and Henry Gopher (Panther clan) used the brand HG.

Seminole cattlemen holding branding Irons. Brighton Seminole Indian Reservation, 1941. Courtesy of Judy Montsdeoca Bronson.

Left to right: Willie Gopher, Sr., Joe Henry Tiger, Jack Smith, Sr., Frank Huff, Sr., Andrew J. Bowers, John Josh, Naha Tiger, Toby Johns, Frank Shore, John Henry Gopher, Lonnie Buck, Charlie Micco, and Harjo Osceola.

Seminole Indian cowboy Charley Micco and grandson Fred Smith on horseback on a cattle ranch. Brighton Reservation, 1950. Courtesy of the State Archives of Florida, image c013676.

Charlie Micco helped establish the cattle management program when the U.S. government shipped a starter herd of starving Hereford cattle from the Dust Bowl states to the Brighton Reservation in 1936. Fred Smith became president of the Seminole Tribe of Florida in 1971.

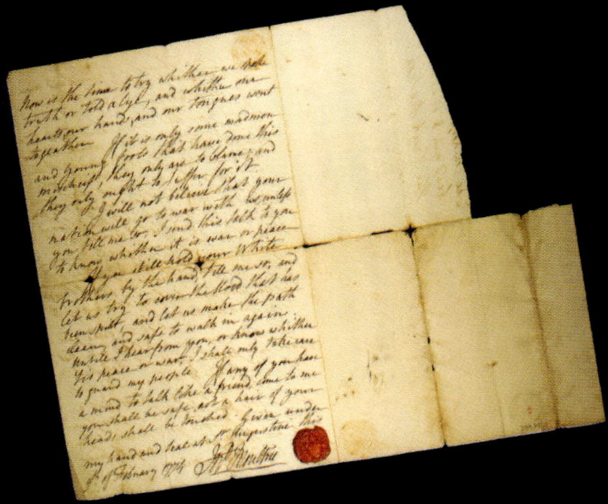

1774 Cowkeeper letter facsimile. Courtesy of Ah-Tah-Thi-Ki Museum, Seminole Tribe of Florida.

The Seminole Tribe of Florida owns a letter to "The Cowkeeper and Other Headmen and Warriors of the Creek Nation" dated February 9, 1774 by John Moultrie, the Lieutenant Governor of the Territory of British East Florida. Written before the term Seminole was commonly used, the letter is significant as the first documentary proof of the identity of the band's leader and their relationship with the British.

Cowkeeper came to Florida with his band of Oconee Indians around 1740 to assist in General Oglethorpe's attack on the Spanish at St. Augustine. Afterwards he established the town of Cuscowilla, on the plains south of present-day Gainesville. There the emerging Seminoles gathered cattle that had strayed from the abandoned missions or old Spanish ranches, eventually creating large herds and ranches envied by white settlers.

The letter arose from a conflict involving 2.1 million acres of land claimed by both the Creeks and the Cherokees. When it appeared to be leading to war between the English and the Indians, the English government inquired whether Cowkeeper's people were preparing for war. Governor Patrick Tonyn met with Cowkeeper and other leaders on March 14, 1774, averting a war until the American Revolution erupted.

Cowkeeper Letter Transcript

Manifesto to the Creek Indians
Gen. Moultrie
1774

Talk to the Cowkeeper and other head men and warriors of the Creek Nation.

Men and Warriors,

I send you bad news. I have heard that some of your nation have struck the great Kings children in Georgia, and spilt their blood upon the ground. The white people grew angry and have killed some of them.

You know that I have always told you that I loved my people, and that I loved your people, and that I wanted to see them live like Brothers, and that our Children should grow up together or on the same land. This is the great King's desire, who will be sorry and angry to hear this bad news.

Let us like good men and wise men, and like men that love our people, force the Hatchet out of the hands of mad men and fools.

When any of you have met me, I have told you that I was your friend and that I loved your people. You told me that you were my friends and loved my people.

Now is the time to try whither wee make truth or told a lye, and whither our hearts, our hands, and our tongues went together. If it is only some madmen and young fools that have done this mischief, they only are to blame; and they only ought to suffer for it.

I will not believe that your nation will go to war with us, unless you tell me so. I send this talk to you to know whither it is war or peace. If you will hold your white brothers by the hand, tell me so, and let us try to cover the blood that has been spilt, and let us make the path clean, and safe to walk in again while I hear from you, or know whither 'tis peace or war. I shall only take care to guard my people. If any of you have a mind to talk like a friend, come to me you shall be safe, not a hair of your head, shall be touched.

Given under my hand and seal at St. Augustine this 9th of February 1774. Gen Moultrie

Manifesto to the Creek Indians

Gov: Moultrie

1774

C

The Cowkeeper & head men & Warrio[rs] the Creek Nation

Talk to the Cowkeeper and other head men and Warriors of the Creek Nation.

Men and Warriors.

I send you bad news. I have heard that some of your nation have struck the great King's Children in Georgia, and spilt their blood upon the ground. the white people grew angry and have killed some of them.

You know that I have always told you that I loved my people, and that I loved your people, and that I wanted to see them live like Brothers, and that our Children should grow up togeather on the same land. This is the great King's desire, who will be sorry and angry to hear this bad news. Let us like good men and wise men, and like men that love our people, force the Hatchet out of the hands of Madmen and fools.

When any of you have met me, I have told you that I was your friend and that I loved your people. You told me that you were my friends and loved my people.

now

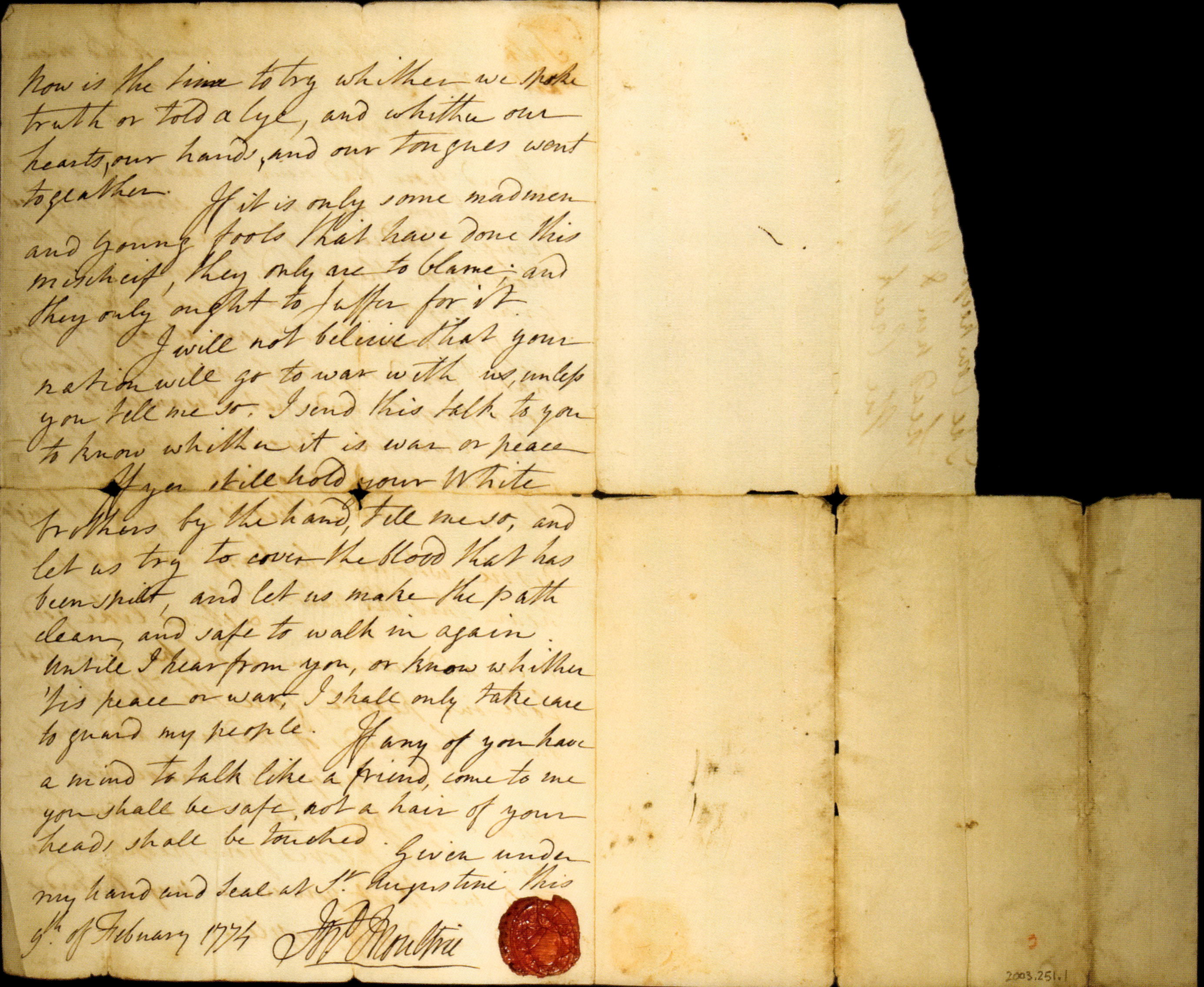
now is the time to try whither we spoke truth or told a lye, and whither our hearts, our hands, and our tongues went togeather. If it is only some madmen and young fools that have done this mischeif, they only are to blame; and they only ought to suffer for it.

I will not believe that your nation will go to war with us, unless you tell me so. I send this talk to you to know whither it is war or peace.

If you still hold your White brothers by the hand, tell me so, and let us try to cover the blood that has been spilt, and let us make the path clean, and safe to walk in again. Untill I hear from you, or know whither 'tis peace or war, I shall only take care to guard my people. If any of you have a mind to talk like a friend, come to me you shall be safe, not a hair of your heads shall be touched.

Given under my hand and Seal at St. Augustine this 9th of February 1774

Jno Moultrie

Moving cattle. Brighton Seminole Indian Reservation. Photo by Carlton Ward, Jr.

Herding cattle into pens on the Brighton Reservation: (left to right) Adam Osceola, Eric Johns, Bobby Yates, and Adam Turtle.

Seminole Indian cowboys marking and branding a calf in the corral during round-up. Brighton Reservation, 1950. Courtesy of the State Archives of Florida, image c014304.

Left to right: Andrew J. Bowers, Joe Henry Tiger, Willie Gopher, Sr.

Rudy Osceola and his grandson, Rowdy. Big Cypress Seminole Indian Reservation. Photo by Carlton Ward, Jr.

Rubber boots are sometimes worn when working in wet pastures.

Contemporary Ranching

Cowmen and Cowwomen

Ranch work is often a family tradition in which many occupational skills are transmitted through generations. People raised on ranches develop "cow sense," a deep understanding of and ability to predict the behavior of cattle. Members of Florida's ranching community are diverse, encompassing men and women, Crackers, Seminoles, African Americans, and Latin Americans.

Crackers have been a dominant force in Florida ranching since the 19th century. Crackers are sometimes defined as the rural white settlers of Celtic heritage who entered Florida from Alabama, Georgia, and the Carolinas. Historically, Cracker culture reflected many traits derived from rural life in Scotland and Ireland: open-range herding; a leisurely lifestyle with an emphasis on hospitality, food, drink, and outdoor sports; and a strong sense of honor, self-reliance, and tenacity.

Some of Florida's earliest African American cowmen were Black Seminoles—free men or runaway slaves who chose to live with the Seminoles. African Americans who migrated to Florida from other states often brought cattle raising skills with them. Since the 19th century, there have been many African American cowmen throughout the state.

Lawrence Silas, 1986. Courtesy of the Osceola County Museum.

The Silases were respected African American cattlemen. Tom Silas worked for a Florida cattleman after the Civil War, then developed a 2,000-acre ranch near Kenansville, and became a trusted friend to many ranchers. His son, Lawrence, won acclaim for his character and knowledge of cattle. He ultimately owned thousands of cattle on ranchlands in Kissimmee and along the St. Johns River. Folklorist Zora Neale Hurston wrote about Lawrence Silas for the September 5, 1942 Saturday Evening Post*, in which he vividly described a stampede during a cattle drive:*

"Never will forget one big stampede. I saw...at Fort Basinger. Eleven hundred steers went into a stampede. We men heard it in time and run in every direction. The stampede headed for a big swamp. Wasn't a thing we could do. The next morning we followed the trail down to the edge of the swamp. We knowed that the ground was too soft for 'em to get across. They didn't; you couldn't see no cows at all. All you could see was horns, just a whole lot of horns."

Bud Adams, Ft. Pierce. Photo by Carlton Ward, Jr.

Alto "Bud" Adams, Jr. is the venerated patriarch of the Adams Ranch, which has properties in St. Lucie, Osceola, and Okeechobee Counties. He established the Braford breed by crossing Brahman and Hereford cattle.

Adams Ranch cowboys herd distinctive red and white Braford cattle, St. Lucie County. Photo by Carlton Ward, Jr.

The Adams ranch is known for its ecological integrity and preservation of natural ecosystems.

Carlton Family

The Carlton family has been influential in Florida since 1843, when Alderman Carlton moved from North Carolina to settle in Florida under the Armed Occupations Act. One of the state's pioneer cattle ranchers, he registered his first cattle brand in 1852. Today, Carlton family members raise cattle and grow citrus on their properties in Hardee, Highlands, De Soto and Okeechobee Counties.

Notable contributions by the Carlton family include Doyle E. Carlton's service as governor of Florida from 1929-1933, during the Great Depression and service by his son, Doyle Carlton, Jr., for three terms as state senator. Doyle Carlton, Jr. and his wife Mildred worked tirelessly to establish the pioneer village at the Florida State Fair, which was named Mildred W. & Doyle E. Carlton Jr. Cracker Country in appreciation of their efforts.

Today, Doyle Carlton III raises cattle and citrus and serves as chairman of the board of directors of the Florida State Fair Authority. He considers it a blessing that the newest generation of the Carlton family—the ninth generation of Floridians—loves the ranching and agricultural life and wants to continue the family tradition of agriculture and land stewardship. To them cattle ranching is more than a business; it is a cherished way of life.

Albert and Martha Carlton with family, Wauchula, ca. 1895. Courtesy of the State Archives of Florida, image FR0470.

Back row L-R: A.H Carlton, C.J. Carlton, A.W. Carlton, Ella Carlton, Thomas N. Carlton, Carl S. Carlton. Front row L-R : G.S. Carlton, Albert Carlton, Leland Carlton, Martha (McEwen) Carlton, and Doyle E. Carlton (later to become governor of Florida).

Three generations of Doyle Carlton on horseback, Hardee County, ca. early 1950s. Photo courtesy of Doyle Carlton III.

L-R: Doyle Carlton, Doyle Carlton, Jr., Doyle Carlton III.

Doyle Carlton III with his son and grandsons, Roman III Ranch, Highlands County, 2011. Photograph by Carlton Ward, Jr.

L-R: Doyle Carlton III, Patrick, Parker and Dale Carlton, Jacob Bolin.

Tom Everett. Marion County, March 2008. Photo by Robert L. Stone. Courtesy of the Florida Folklife Program/Florida Department of State.

Thomas Everett, Sr. (1932–) worked cattle throughout Central Florida for nearly six decades until a horse accident disabled him.

Florida ranching traditions have been influenced by Hispanic culture from Spanish and Minorcan colonists to recent residents. Many Latin Americans who arrived in the late 20th century preserve ranching traditions, though few actively work as ranchers or cowboys. On numerous small ranches in the rural Redland area southwest of Miami, Latin Americans keep horses or recreate the architecture, customs, and leisure activities of their homelands. Many participate regularly in rodeo events within the Latin community.

Women assume many roles, performing all but the most dangerous and strenuous work. It is not uncommon for them to work cattle from horseback, operate tractors, bale hay, or run errands such as buying feed, veterinary supplies, truck and tractor parts, hardware, and other provisions. Women often maintain the books, manage finances, track political developments that affect the ranch or cattle industry, and are active in youth activities such as 4-H and FFA. Women's auxiliary organizations, such as Florida Cattlewomen, make classroom presentations to inform students about the beef industry. Since the extended matrilineal family is an important social unit, some Seminole women own cattle herds, participate in tribal cattle management, and work at the weighing pens during the annual roundup.

José Carlos Martinez. Miami-Dade County, April 2008. Photo by Robert L. Stone, courtesy of the State Archives of Florida, image FS060815.

José Carlos Martinez is a cowboy, rodeo champ, and horse trainer from Pinar del Rio, Cuba. In Florida he has worked as a cowboy and horse trainer at ranches in southwest Miami-Dade County.

Lewis Clayton. Gainesville, June 2006. Photo by Robert L. Stone, courtesy of the State Archives of Florida, image FS060818.

Lewis Clayton raised crops and hogs before he began cattle ranching on Kanapaha Prairie, Gainesville.

"Hombres." Arcadia, 2008. Photo by Jimmy Peters.

Mexican American cowboys at the VC Ranch: left to right, Alfredo Manriguez, Lynn Hollingsworth Mills (daughter of ranch owner Victor Clyde Hollingsworth), Francisco Barbosa, Javier Manriquez, Luis Manriquez, Teofilo Manriguez, and Pancho "Juan" Martinez.

Iris Wall with granddaughter Whitney Edwards. Indiantown, September 2006. Photo by Robert L. Stone. Courtesy of the Florida Folklife Program/Florida Department of State.

2006 Woman of the Year in Agriculture, Iris Wall owns the High Horse Ranch in Indiantown. Her granddaughter, Whitney Edwards, a hairdresser, often helps her with cattle work.

Partin Family sitting on fence. Osceola County, ca. 1952. Courtesy of the Osceola County Museum.

The Partins of Osceola County are a well known cattle ranching family. Along the fence, left to right: Richard Earl Partin holding Janie Partin Pickering; Dorothy Platt Partin; Barbara Partin, Bertha Bass Partin (Mrs. H.O.); Kathy Tyson Baker, Connie Autrey Partin, Judith Partin; Michael Steven Partin; "Peetie" Partin; Martha Thomas Booth; Becky Partin Kemfer; Henry Hyatt Partin, Jr.; Susan Bronson Williams, Steve Partin; Katherine Partin and Sherry Bronson. Standing, L-R: Henry O. Partin, "Geech," "Doc," and Doug Partin.

RRR Ranch. Medley, April 2008. Photo by Robert L. Stone, courtesy of the State Archives of Florida, image FS060813

A cigar clenched in his teeth, a Cuban American cowboy attempts to rope a steer at the RRR Ranch.

Renee Strickland. Parrish, 2007. Photo by Tina Bucuvalas. Courtesy of the Florida Folklife Program/Florida Department of State.

On the Strickland Ranch in Parrish, Renee Strickland shares the work with her husband, Jim. The Stricklands engage in the international trade of Florida cattle.

Ranch Life

Ranch life revolves around the annual cycle of work generated by the cattle. Most Florida cattle ranchers are in the cow-calf business. On many ranches calves are generally born in the fall and spring. At two to four months they are wormed, vaccinated, earmarked, and the males are castrated. When the calves are 7 to 9 months old they are weaned and sold at weights ranging from 300 to 600 pounds. Since Florida's environment does not support efficient grain production or highly nutritious forages, calves are shipped to western states where they are grazed and fed until they reach finish weights.

Cattlemen refer to gathering livestock and performing procedures as "working" the cattle. Most cattle require working twice a year. On large ranches cattle are worked year round; smaller herds may require working only every two months. The cattle are herded into pens using a system of gates and chutes to separate the calves, brood cows, bulls and steers. Calves are usually worked in a mechanical "calf table," a steel frame that holds them in position while multiple procedures are performed. Larger animals are worked in mechanical-hydraulic squeeze chutes. Most Seminole cattle belong to small family-owned herds tended by a central crew under a cooperative arrangement. They often work the cattle by roping and using manpower to handle the stock because cow pens, calf tables and squeeze chutes may be far away.

The annual roundup and shipping is the busiest time on the ranch, and the major annual payday. Only the largest operations employ a full-time crew of six or more cowboys year round. When ranchers need additional help to work cattle they may hire "day-riders," or receive assistance from family members, friends, or neighboring ranchers with whom they have a reciprocal agreement.

Working cattle. Vero Beach, September 2008.
Photo by Bob Montanaro.

Cowboys herd heifers into Treasure Hammock Ranch pens where they will be treated with insecticide for horn flies. Left to right: Will Barker, Rob Tripson, Daniel Brolman, Sean Sexton (on tractor), Davey Pitman.

Working a cow in squeeze chute. High Horse Ranch, Indiantown, September 2006. Photo by Robert L. Stone. Courtesy of the Florida Folklife Program/Florida Department of State.

Left to right: Cab Butts sprays the cow's back with an external parasite treatment, Whitney Edwards vaccinates for respiratory disease, Marshall Davis and Jim Harvey administer two different oral medications for internal parasites.

Mike Wilder (1954–). Working cowboy saddle, leather. Rain slicker, rope, and cow whip attached. Saddle, slicker and rope loaned by Mike Wilder. Whip made and loaned by Ned Waters. Photo by Ray Stanyard.

This saddle and attached accessories are typical of those used by working cowboys. The saddle itself is designed for comfort and durability, was custom-made to fit the rider and his horse, and is devoid of frills and fancy decoration.

The full-length yellow vinyl-reinforced nylon rain slicker is split at the waist to allow it to drape over the horse and saddle, as a wet saddle will quickly chafe the rider.

The 14-foot nylon cow whip is a popular length for working from horseback; long enough to afford adequate reach, but not so long as to be difficult to handle.

The nylon rope may be used to lasso cattle which range from small calves to bulls weighing up to 2,000 pounds.

Many cowboys also attach small saddle bags, or pockets, (not displayed here) to carry a variety of items.

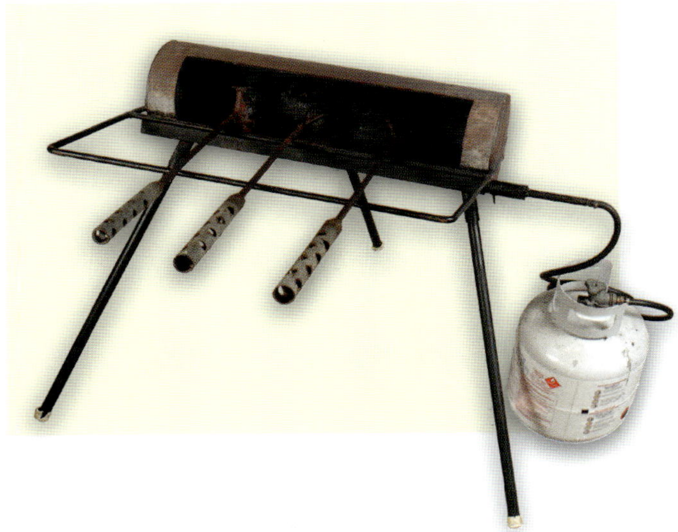

Work Area. Photos by Robert L. Stone.

Work table. Cypress wood. Built and loaned by Sean Sexton.
The table contains items typical of those used when working cattle in a squeeze chute. All items loaned by Sean Sexton:

1. Antibiotic, LA-200 brand. Brown glass bottle.

The antibiotic is administered with small syringes (not displayed) to treat cattle for infections. Alcohol and cotton balls are used to clean the area.

2. De-horner. Steel.

Cattle are de-horned to protect themselves and those who handle them from injury. After de-horning, the area is treated with blood-stop powder.

3. Notebook.

Cattlemen keep detailed records on each animal.

4. Nose leads. Steel, with rope attached.

Nose leads are used to control cattle resistant to being worked.

5. Large syringe. Stainless steel.

This is an older device used for administering oral medicine, such as de-wormer. The modern method employs an applicator gun, such as the one being used by two men in the photo (page 57), connected by a plastic hose to a one gallon container as shown on the table.

6. Large red marker crayon.

The crayon is used for temporary identification of cattle as they are worked.

Branding iron set. Steel. Loaned by Sean Sexton.

This set of branding irons consists of 0 through 9 and the "Bar G" ranch, or holding, brand. Cattlemen use the numbered brands for year designation, and some may give each animal an individual identification number. The holding brand identifies the ranch or owner.

Branding iron heater pot. Steel. Loaned by the Florida Cattlemen's Association.

A portable propane-powered gas-fired "pot" like this is probably the most common method of heating branding irons today. However, some ranches use electrically–heated brands.

Barbed wire, Adams Ranch.
Photo by Carlton Ward, Jr.

Fence maintenance is a continuing task on cattle ranches.

Uploading. Sarasota, 2008. Photo by Jimmy Peters.

Arthur Kersey loads cattle into a tractor trailer at the High Hat Ranch.

Cowboy with calf. St. Lucie County. Photo by Carlton J. Ward, Jr.

Sean Moss carries a tired calf to help it keep up with the herd as they move across the Adams Ranch.

Working cowboys with cow-dogs. Okeechobee County, January 2008. Photo by Robert L. Stone, courtesy of the State Archives of Florida, image FS060805.

Left to right: Cab Butts and Ernie Carrier head out with a pair of black mouth yellow cur dogs to gather cattle early on a chilly January morning at the Dixie Ranch.

Occupational Skills

The many occupational skills of the working cowboy are learned informally, through observation and imitation. In the past, roping cattle was not a common practice in Florida because of the extensive areas of dense scrub where a snagged rope could result in a serious accident. Since screw-worm treatment required that each animal be roped and subdued, cowmen adopted lariats and replaced hornless McClellan saddles with western-style saddles. Today, roping is a valuable occupational skill.

Traditional methods of cattle identification include flank branding with a hot iron and "marking" one or both ears using a knife to cut a combination of standardized shapes. It is common practice to mark and brand cattle because earmarks are often easier to read from a distance. One owner may have several different brands and earmarks, and individual family members often have their own brands and marks. Today, ranchers usually brand by applying an iron heated by a portable "pot" fueled by a small propane tank or by using an electrically heated branding iron. Some use numbered plastic ear tags rather than earmarks.

Tending cattle often requires that certain animals, such as cow and calf pairs, are sorted from the herd and separated. Efficiently sorting stock from horseback is one hallmark of a highly skilled cowboy. This complex task requires excellent horsemanship, a well-trained horse, and plenty of cow sense. Conversely, the ability to match up a cow with her calf after they have been separated is known as "mammying-up." This unusual skill requires a photographic memory and the ability to observe subtle cattle behavior. Those who excel at these skills are highly regarded by their peers.

Good equestrian skills and a well-trained horse are essential to successful cattle tending and serve as important symbols of ranching culture. Ranch horses are trained for specific tasks, such as stock sorting and roping. Although the American Quarter Horse is used by cowboys throughout the U.S., some Floridians prefer the smaller Cracker Horse for its agility and stamina during hot weather.

Roping calf. Brighton Seminole Indian Reservation. January 2008. Photo by Robert L. Stone, courtesy of the State Archives of Florida, image FS060839.

Koty Lewis attempts to lasso the hind legs of a calf that Alex Johns has roped by the head. Most Seminole cattle are owned by 66 families. They are managed in a cooperative arrangement and squeeze chutes are not available for each herd. Consequently, the Seminole cattle crews do a lot of roping.

Cowboy with whip. May 2008. Photo by Carlton J. Ward, Jr.

Matt Harrison cracks a cow whip as he moves a herd. A fourth-generation cowboy, he runs the Harrison Cattle Company with his father on land in Hardee, De Soto, Sarasota, and Manatee Counties.

Sean Sexton roping a calf at Treasure Hammock Ranch. Vero Beach, October 2008. Photo by Bob Montanaro.

Sean Sexton works from the ground to rope a calf in order to fit it with an ear tag as its mother watches protectively.

Dally and hard-and-fast tie. Kissimmee, October 2008. Photos by Robert L. Stone, courtesy of the Florida Folklife Program/ Florida Department of State.

Many Florida cowboys fasten the rope to the saddle horn using a knot to make a "hard-and-fast tie." This contrasts to the western "dally" method of wrapping the rope around the horn, which allows the rope to be played in or out. Billy Bellamy demonstrates both techniques.

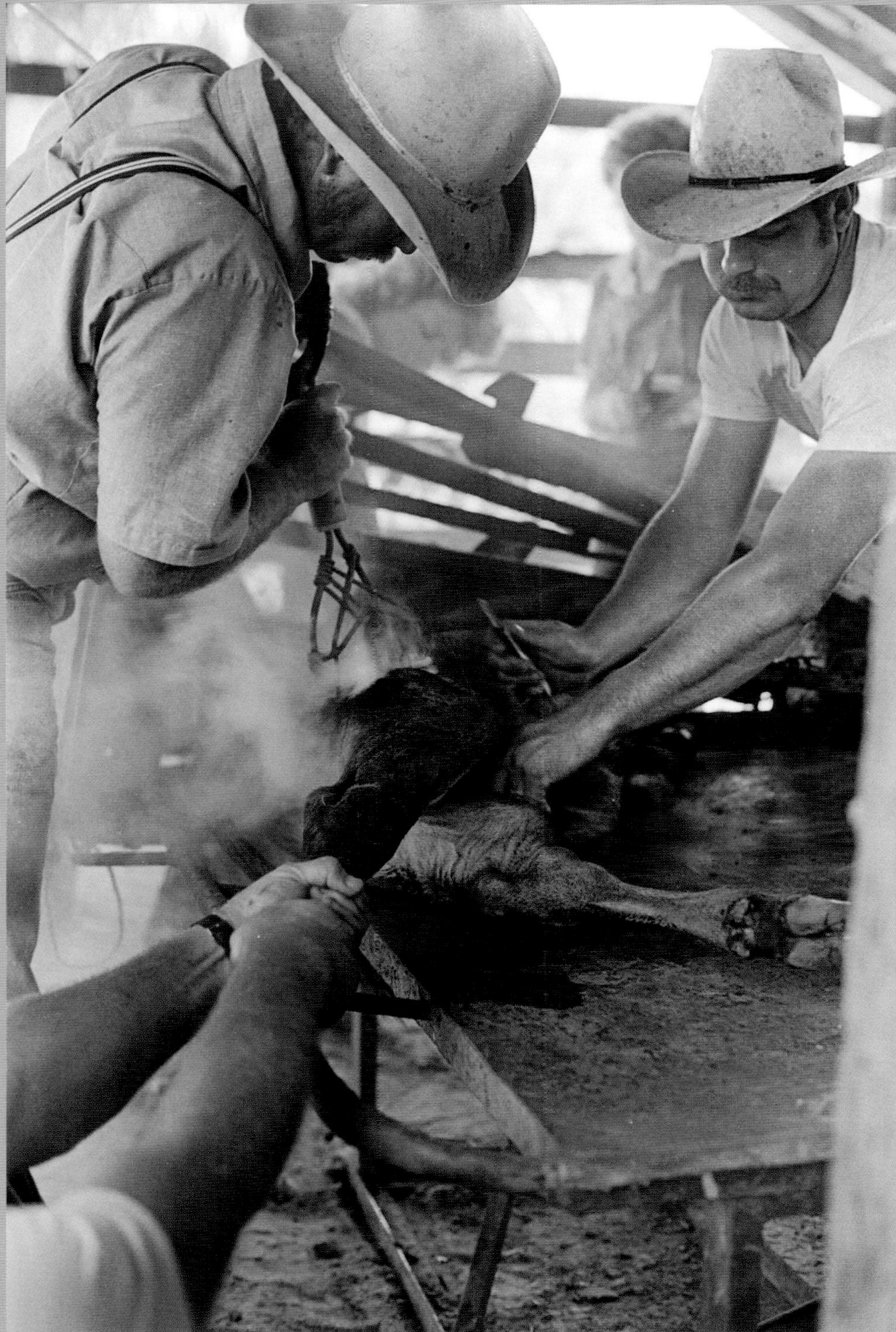

Terry Myers and Ronnie Sylvester castrating, marking, and branding cattle at Buck Island Ranch. Lake Placid, 1984. Photo by Blanton Owen. Florida State Archives.

"Cut it Out." Arcadia, 2008. Photo by Jimmy Peters.

Working as a skilled team, Frank Langsford and his ranch horse expertly cut a cow from the herd at the VC Ranch.

Graphic: Earmarks. From 1984 Florida Folk Festival program booklet (White Springs: Florida Folklife Program), page 12. Courtesy of the Florida Folklife Program/ Florida Department of State.

Reading Ear Marks

Ear marks are usually "read" from left to right as you face the animal. Seasoned cowmen say marks have advantages over brands. When you approach a cow, it will invariably turn its head toward you. Properly notched ears are clearly visibile, making identification quick and easy. In addition, if a brand is mud-covered or old, or if the cow is too far away for a brand to read, the ear marks provide identification. The diagrams below show wha: the different marks or cuts are called:

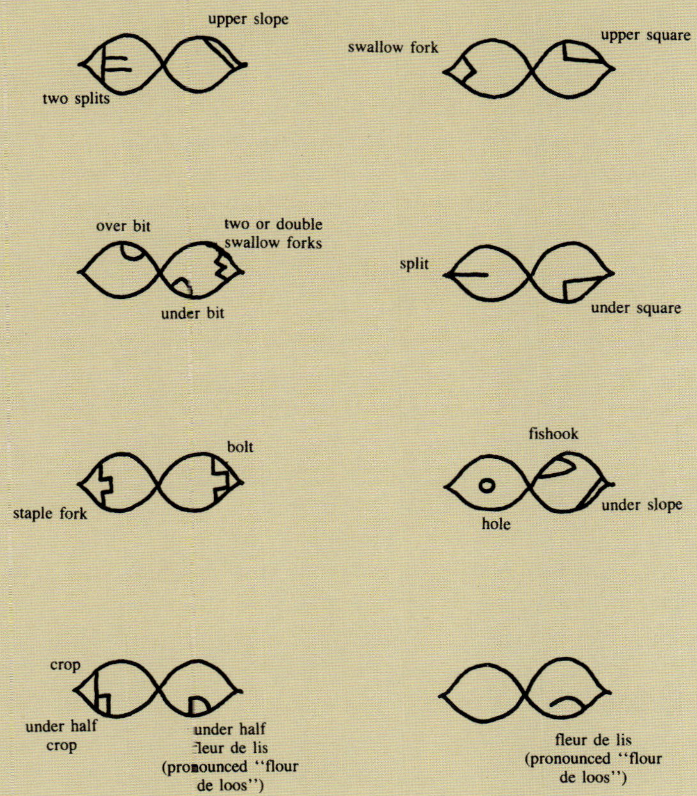

From field notes by folklorist Blanton Owen, 1984.

Earmarking a calf. Brighton Seminole Indian Reservation, January 2008. Photo by Robert L. Stone, courtesy of the Florida Folklife Program/Florida Department of State.

Bobby Yates cuts earmarks with a razor-sharp pocket knife while Carlos Leon holds the calf down.

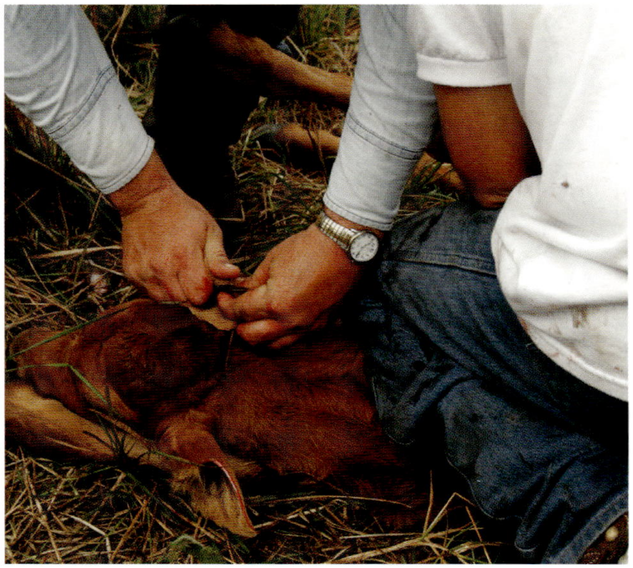

Customs

Traditional customs are embedded in the daily life and special events of Florida's ranching community. Foodways, for example, often reflect southern roots. Rice and tomato gravy, cornbread, catfish and hushpuppies, okra, beans, barbeque, and swamp cabbage are favorite dishes. Swamp cabbage, called *taal-holelke* by the Seminoles, is a symbol of cultural identity.

Food plays an important role during events such as the annual roundup and shipping of cattle at the Brighton and Big Cypress Seminole Indian Reservations. For two weeks during July and August, the herds of six or seven families are rounded up daily, transported in trailers to pens where they are weighed and tallied, then loaded into cattle trucks and shipped to western feed lots. On the day their cattle are shipped, each family prepares and serves a buffet-style noon meal for about 150 workers and family members. Most serve dishes such as fried chicken and macaroni and cheese, but some women prepare Indian fry bread or swamp cabbage. At the mealtime gathering, friends and family from both reservations share food and conversation.

Many members of Florida's cattle ranching community are devout Christians. Worship and prayer is an integral part of family life, rites of passage, public events, and community meetings. At rodeos, opening ceremonies include a Christian prayer, and contestants may pray prior to competing. "Cowboy church" is a popular form of worship tailored to cowboys, ranchers, and members of agricultural communities. Prayers may reference horses, cattle, and ranch work. Worship services are held at rodeo grounds or other outdoor facilities. Rodeos often provide Sunday morning cowboy church for traveling contestants, and contestants commonly include Christian or cowboy church motifs on their apparel and equipment.

Rites of passage, which include observances of births, marriages, death, and other major life experiences, often express the traditional culture of their participants. When a member dies, the ranching community may express regard for the deceased and make statements of identity and cultural values through funeral rituals. Obituaries of cattlemen appear in the Florida Cattlemen's Association magazine.

Buddy Mills' son, Chad, with prepared swamp cabbage. Okeechobee, July 2007. Photo by Robert L. Stone, courtesy of the State Archives of Florida, image FS060826.

Chad Mills used an old family recipe to prepare this pot of swamp cabbage for the National Day of the American Cowboy celebration.

Buddy Mills cutting cabbage. Okeechobee County, December 2006. Photo by William Mansfield, courtesy of the Florida Folklife Program/Florida Department of State.

Buddy Mills will make swamp cabbage from the heart of palm he has cut from a cabbage palm tree. He modified his ax for this purpose by welding on additional metal to widen the blade.

Barefoot cowboy with coffee pot. Osceola County. Courtesy of the Osceola County Historical Society.

A cowboy starts his day with hot coffee prepared on a campfire.

Seminole round-up lunch. Big Cypress Seminole Indian Reservation, July 2007. Photo by Robert L. Stone, courtesy of the State Archives of Florida, image FS060822.

Under the shade of a thatched chickee, cowboys and other workers share a lunch provided by the cattle owners whose herds are being rounded up and shipped. Men in light cowboy hats (left to right): Joe Osceola, Sr., Alex Johns, Paul Bowers, Wilse Bruised Head. Woman right foreground: Mary Jene Koenes. Man in camouflage ball cap and white tee shirt, right: Earl Kirkland. Others unidentified.

VIDA Ranch barbecue, Osceola County, November 2007. Photo by Robert L. Stone, courtesy of the State Archives of Florida, image FS060816.

Norman West cooks country sausage links on a homemade barbecue grill for the VIDA Ranch crew lunch.

Baptism at Cowboy Church. Williston Horseman's Park, October 2008. Photo by Robert L. Stone, courtesy of the State Archives of Florida, image FS060828.

Pastor Billy Keith (right) of Cross Brand Cowboy Church baptizes Amber Thompson in a cattle trough.

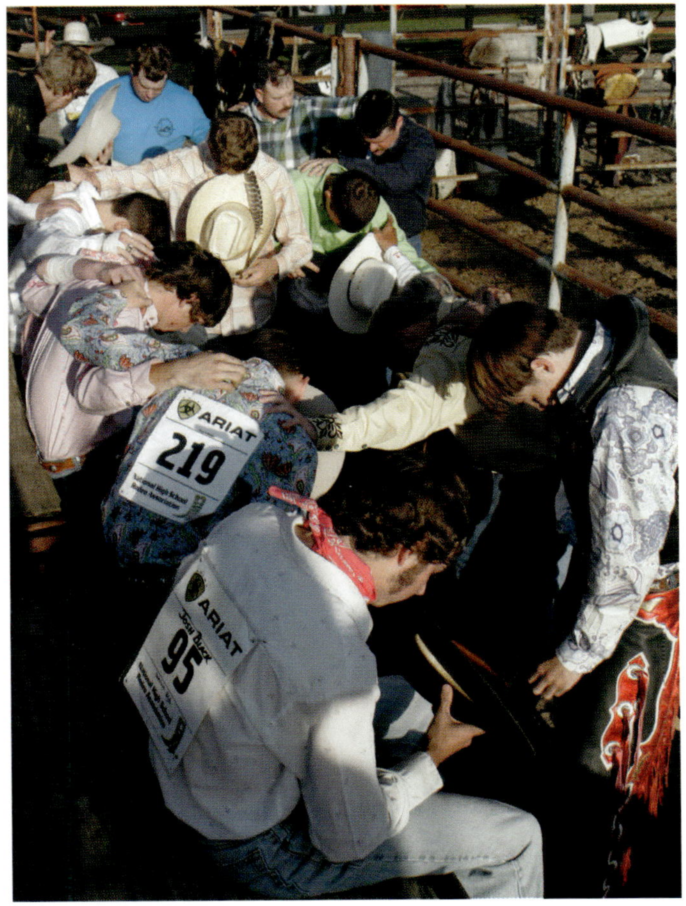

Praying Williston High School Rodeo contestants. Williston Horseman's Park, April 2007. Photo by Robert L. Stone, courtesy of the State Archives of Florida, image FS060833.

Praying bull riders. Williston Horseman's Park, March 2007. Photo by Robert L. Stone, courtesy of the State Archives of Florida, image FS060803.

Stephen "Preacher" Keighley leads fellow bull riders in an extended prayer behind the chutes at the annual Bull Mania.

Coffin in horse trailer
at Junior Mills' funeral.
Basinger, June 2006.
Photo by David Chandler.

Coffin in front of horse trailer and horse at George "Junior" Mills' funeral. Agri-Civic Center, Okeechobee, June 2006. Photo by David Chandler.

For the funeral for venerated Okeechobee Cracker cowman George "Junior" Mills, his "Rocking Double X" brand was burned into the cedar casket. Spur, the horse he rode for more than 20 years, stands behind the casket.

North Florida Livestock Market. Ellisville, January 2009. Photo by Robert L. Stone, courtesy of the Florida Folklife Program/Florida Department of State.

A buyer (foreground, white hat) signals his bid by raising his hand in response to the patter of auctioneer Aubrey Bailey (background center). The calf's weight of 425 pounds is displayed above the auctioneer.

Auctions

Auctions are a traditional setting for buying and selling cattle and ranch horses. Cattle auctions are held once or twice weekly at several locations throughout the state. The Okeechobee Livestock Market is one of the largest east of Texas. At the typical livestock market, cattle are directed to an auction ring surrounded by seated buyers. As the auctioneer solicits bids through his rapid patter, the animal's weight is displayed on a large screen. Bidders communicate by hand signals, the auctioneer announces the number of the winning buyer, and the animal exits the auction ring. An animal is often in the auction ring less than ten seconds. Livestock markets also provide an opportunity for networking and socializing among the cattlemen. In recent years cattle sales through Internet auctions, satellite television, and video have become increasingly popular with ranchers who own large herds of uniform calves. Since most Florida cattle are owned by small-scale ranchers, there remains a need for livestock markets where cattle are auctioned individually.

Buyers' pens, Okeechobee Livestock Market. Okeechobee, May 2004. Photo by Robert L. Stone courtesy of the Florida Folklife Program/Florida Department of State.

About 2,500 head of cattle per week are auctioned, one at a time, at the Okeechobee Livestock Market. The cattle are then sorted into pens designated for each buyer. The day after an auction, the pens are emptied and the cattle are loaded into trucks and shipped.

Cow-Dogs

Florida is one of the leading states in the use of dogs for cattle ranching. Florida cow-dogs are mixed breeds. Probably the most popular breed in the mix is the Southern Blackmouth Yellow Cur. Another popular breed is the Catahoula Leopard, which may have its origins among the French or Native Americans of Louisiana.

Florida cow-dogs perform three important functions. The cow-dog flushes strays from hammocks, scrub and swamps, working easily in areas that are difficult for horse and rider to penetrate. While mounted cowmen drive the cattle from the rear, cow-dogs control their movements by working the sides and front to keep the stock in a cohesive herd. And, since cattle view the dogs as predators, they will seek safety by gathering into a tight group when the cow-dogs repeatedly circle or "ring" them, while barking and nipping.

Cow-dog circling herd. Morriston, June 2006. Photo by Robert L. Stone, courtesy of the Florida Folklife Program/Florida Department of State.

One of Billy Bellamy's yellow curs circles a herd at his Cedar Hill Ranch to keep the cows in a tight bunch. A few good dogs can maintain a large herd tightly grouped for hours if necessary.

Working cow-dog. Okeechobee County, January 2008. Photo by Robert L. Stone, courtesy of the Florida Folklife Program/Florida Department of State.

A yellow cur cow-dog aggressively confronts a cow that has broken from the herd on the Dixie Ranch.

Material Culture

Floridians hand craft specialized equipment used by the cattle ranching community. Mexican vaquero culture played a profound role in shaping the fancy, highly-decorated cowboy equipment, tack and apparel that developed in the West—but there was little vaquero influence in Florida. Most gear used by Florida cowboys is unadorned in comparison to that used in the West.

The cow whip is an essential part of working cowboy gear and an icon of Florida cattle ranching. In the past, cow whips were plaited from tanned buckskin. Today, most Florida cowboys use weather- and rodent-proof whips plaited from braided nylon cord. The Florida cow whip is fastened to a wooden handle by two thongs, a distinctive feature that gives it more flexibility than whips plaited over the handle.

A comfortable, durable saddle is essential. A custom-built saddle fits the rider and horse properly, is designed to meet the demands of its intended use, and lasts for many years.

Ranch brands were used by early settlers and Indians to identify livestock. The practice was especially important when cattle ranged freely. Ranch brands are usually passed to the next generation, and old branding irons become treasured heirlooms. Many brands consist of simple block numbers or letters, a characteristic derived from the British ranching heritage.

Ranch architecture includes entryways, gates, and signs. These utilitarian art forms range from simple hand-made wood and metal components to elaborate designs cut from metal using high-tech computer-controlled plasma arc machines. Designs often incorporate the ranch brand and visual elements, such as cattle, herons, live oaks, and cabbage palms, which reflect the ranch environment.

Riders use spurs to help control the movement of horses. Used properly, they are necessary tools for the protection of the rider, cattle, and the horse itself. Ready-made spurs can be purchased, but many prefer spurs that are custom made to personal specifications.

Chaps are protective leather garments worn over jeans. Chinks are short chaps that have become increasingly popular among Florida cowboys, in part because they are not as hot as full-length chaps. In Florida, chaps are most often seen as colorful, fancy apparel worn by drill teams, bronc and bull riders, and rodeo queens.

George Altman. Wauchula, March 2007. Photo by Robert L. Stone, courtesy of the Florida Folklife Program/Florida Department of State.

George Altman is one of the few who still plait cow whips from buckskin hides he tans himself.

Junior Mills. Okeechobee, June 2003. Photo by Robert L. Stone, courtesy of the Florida Folklife Program/Florida Department of State.

George "Junior" Mills made buckskin whips to supplement his income as a working cowboy. He treated the finished whips with a heated mixture of pine rosin, beeswax, and beef tallow to preserve the leather and prevent barn rats from eating them.

William Noah Waters (1871–1944). Cow whip. Buckskin with guava wood handle. Bartow, ca. 1940. Loaned by Ned Waters. Photo by Ray Stanyard.

Ned Waters inherited this 12-foot whip from his grandfather. It probably was originally a 14 to16-foot whip, but was shortened when it was fitted with a new tail. The hide was tanned using animal brains.

George (1912–2006) and Buddy Mills (1958–). Cow whip, buckskin. Okeechobee, 1975. Loaned by Buddy Mills. Photo by Ray Stanyard.

George "Junior" Mills cut the "strings" for this 11-foot whip then taught his son, Buddy, how to plait the whip. Buddy has continued to craft traditional buckskin whips using the methods he learned from his father. The dark coloration is the result of dipping the finished whip in a heated mixture of beef tallow, beeswax and pine rosin to preserve it and discourage rodents from eating it.

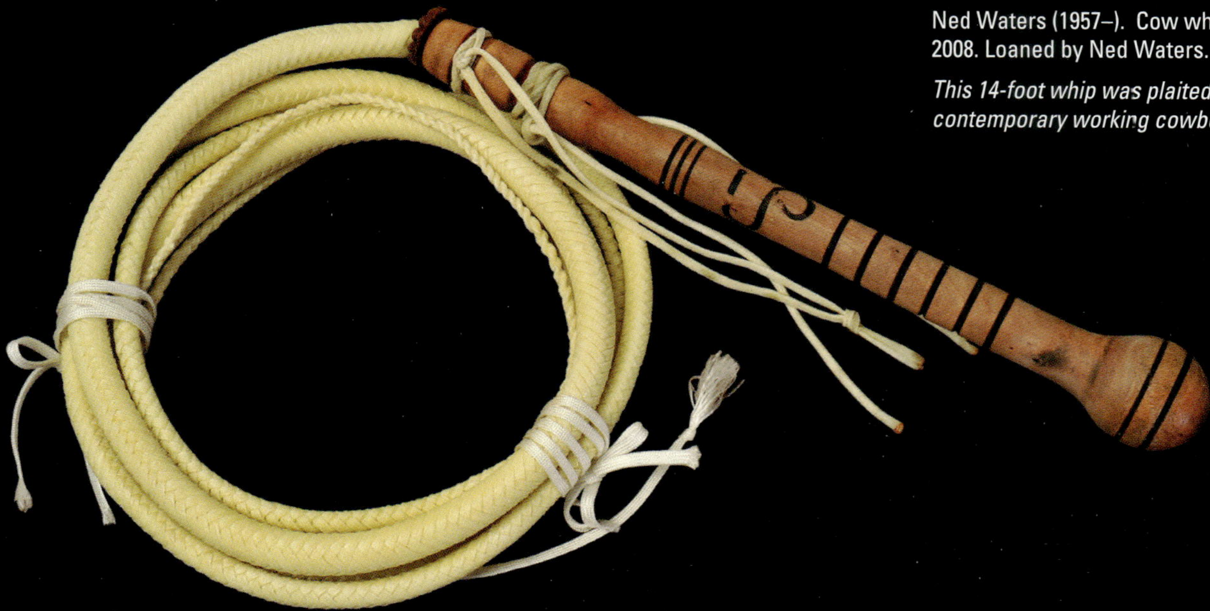

Ned Waters (1957–). Cow whip, nylon with guava wood handle. Bartow, 2008. Loaned by Ned Waters. Photo by Ray Stanyard.

This 14-foot whip was plaited from nylon cord, the material preferred by contemporary working cowboys because it is rot- and rodent-resistant.

George Altman (1946–) Cow whip, buckskin with hickory wood handle. Wauchula 2005. Loaned by the Florida Cattlemen's Association.

This 14-foot whip was plaited from a hide Altman tanned in an alum solution. All Florida cow whips, whether nylon or leather, are loosely fastened to the handle by two thongs, which makes them flexible and easy to use.

Curly Dekle demonstrating whip handling skills at the Florida Folk Festival. White Springs, 1988. Photo by Gregory Hansen, courtesy of the State Archives of Florida, image fa4792.

J. P. "Curly" Dekle crafted buckskin leather whips and was a skilled whip artist. He was a perennial favorite at the Florida Folk Festival, where his presentations included plenty of down-home humor as well as engaging demonstrations of his prowess with whips.

Branding iron maker. Okeechobee, October 2006. Photo by William Mansfield, courtesy of the Florida Folklife Program/Florida Department of State.

Tony Hapner welds a branding iron at Domer's, Inc., a welding and machine shop that has made branding irons since 1926.

Saddlemaker Mike Wilder. Kenansville, September 2006. Photo by Robert L. Stone, courtesy of the Florida Folklife Program/ Florida Department of State.

Seventh-generation Floridian Mike Wilder has earned a reputation for making well-designed saddles that will last for years. He made this commemorative saddle for the Great Florida Cattle Drive of 2006.

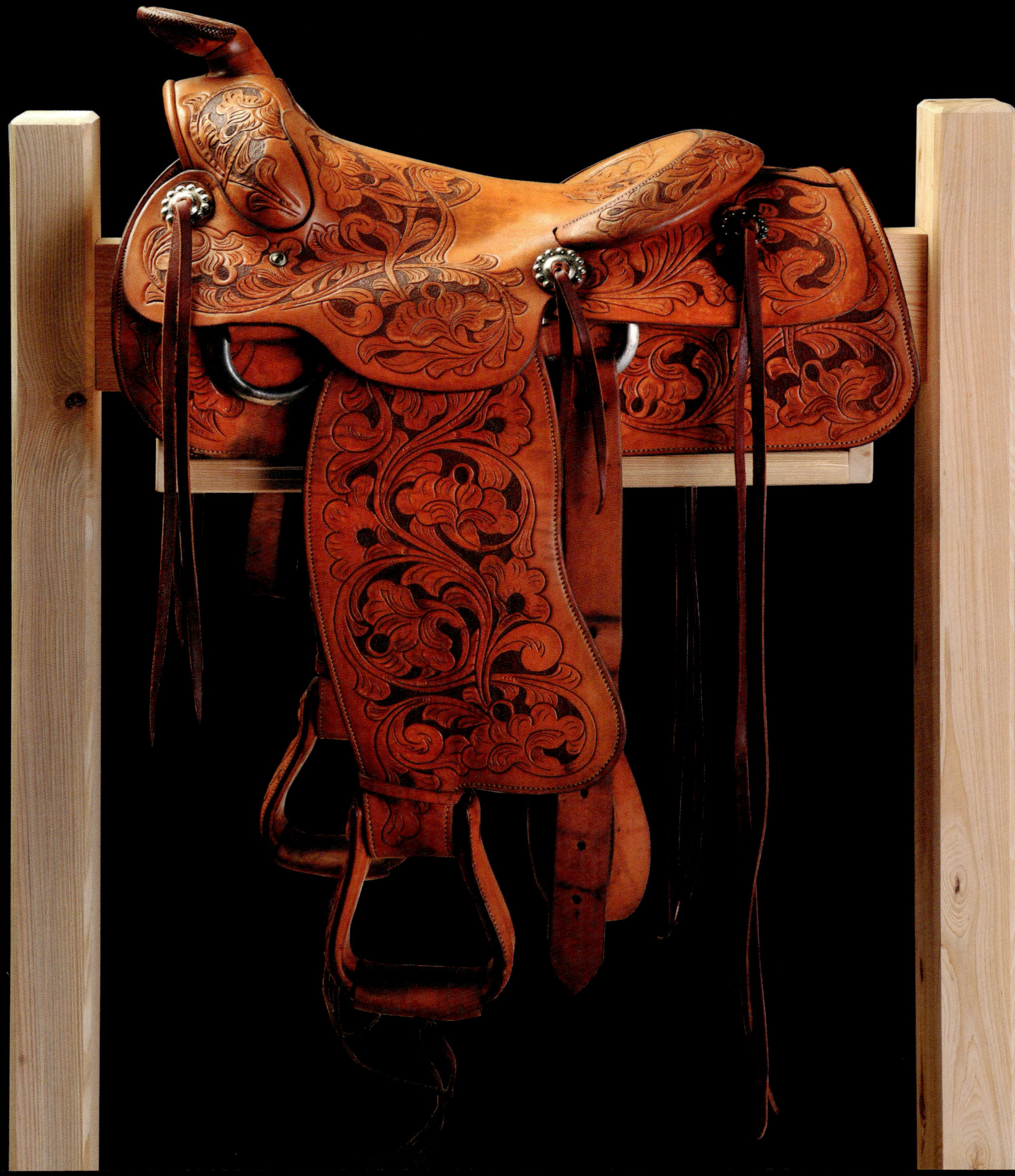

Felix T. Torres (1913–1995). Saddle carved for his son, Felix G. Torres. Miami, 1964. Leather, silver. Loaned by Felix G. Torres. Photo by Ray Stanyard.

At the Torres Saddle Shop in Hialeah, the Torres family has made and repaired saddles and other custom leather goods for several decades. The late Felix T. Torres was from the ranching area of Santa Clara in the province of Las Villas. He moved to Havana, where he became an accomplished saddlemaker. He passed the tradition to his son, Felix G. Torres, when he was very young. Torres saddles enjoy an international reputation for excellence. One of the family's old photos shows the Governor of Havana riding on a Torres saddle.

Felix T. Torres making a saddle with wife Maria E. Torres, son Felix G. Torres and Nery Perez looking on. Havana, Cuba, 1956. Courtesy of Felix G Torres.

George Platt. Okeechobee, January 2008. Photo by Robert L. Stone, courtesy of the Florida Folklife Program/Florida Department of State.

George Platt makes tack and a variety of leather goods for working cowboys. He also builds and repairs saddles.

Babcock Ranch gates. Charlotte County, September 2006. Photo by Robert L. Stone, courtesy of the Florida Folklife Program/Florida Department of State.

The ranch's Crescent B brand is displayed in this unique ranch entryway design fashioned from native limestone, metal and wood.

Bronson Family Ranch gates. Osceola County, November 2006. Photo by William Mansfield, courtesy of the Florida Folklife Program/Florida Department of State.

A painting of a Brahman bull and monogrammed gates adorn this ranch entryway.

Double M. Ranch gates. Zolfo Springs, March 2007. Photo by Robert L. Stone, courtesy of the Florida Folklife Program/Florida Department of State.

The ranch name and cattle figures were cut from steel.

Jeremy Evans. Fellowship, February 2007. Photo by Robert L. Stone, courtesy of the Florida Folklife Program/Florida Department of State.

Jeremy Evans and his wife, Connie, fashion custom spurs in a shop behind their home in Fellowship, west of Ocala.

Jeremy and Connie Evans. Spurs, custom handmade "Gal-Leg" style. Steel with sterling silver and copper overlay and brass rowel. Fellowship, 2008. Loaned by Jeremy and Connie Evans. Photo by Ray Stanyard.

Jeremy made the body of the spur and his wife, Connie, did the overlay and engraving. The rowel is a brass clock gear.

Billy Davis in his spur making shop. Kenansville, April 2004. Photo by Robert L. Stone, courtesy of the State Archives of Florida, image FS060849.

Billy Davis clamps the heel ridge portion of a spur onto a jig in preparation for welding the shank to it. He crafts custom spurs and spurs similar to the classic designs of 1840-1940.

Doyle Rigdon's lace-up boot with spurs. Brighton, September 2008. Photo by Robert L. Stone, courtesy of the State Archives of Florida, image FS060829.

Like many Florida working cowboys, Doyle Rigdon prefers to wear lace-up boots because the snug fit and low heel make it easier to move quickly when working on the ground.

Poppy Moe with chaps and chinks. Ocala, May 2007. Photo by Robert L. Stone, courtesy of the Florida Folklife Program/ Florida Department of State.

Leather worker Poppy Moe specializes in chaps and chinks (short chaps).

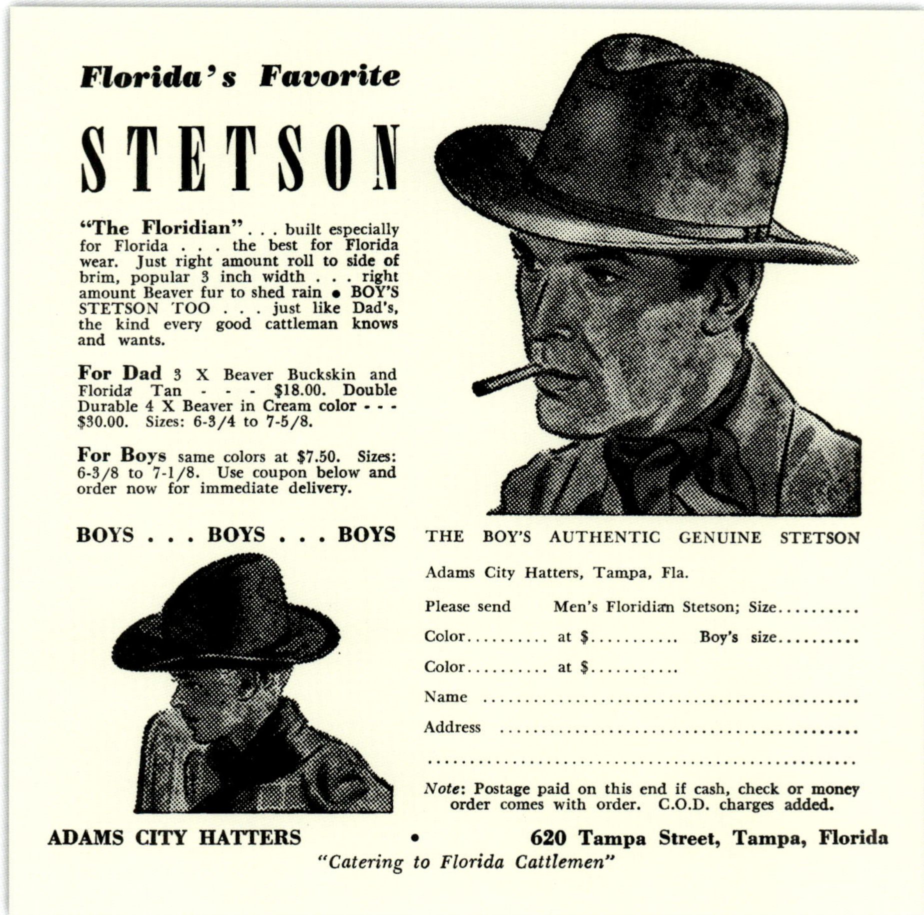

Advertisement for Stetson "The Floridian" hat. *The Florida Cattleman and Livestock Journal*, October 1949.

Perry McLelland. Carved leather fender and stirrup. Lake Worth, 1990. Loaned by HistoryMiami. Photo by Ray Stanyard.

The McLelland family has been in the saddle business for five generations. Jim McLelland became a saddlemaker after returning from World War II. He moved his family to Lake Worth in 1958. His children, Perry, Ted, and Debbie, and Ted Jr. continue to provide fine saddles to cowboys and pleasure riders in south Florida.

Oral Traditions

Cowboy poetry borrows from such sources as folk poetry and song, popular poetry, historical writing, fiction, film and television. Florida cowboy poets express a wide range of themes through verse. Poems may vividly describe the rugged and beautiful landscape, recount memorable events, or express the excitement and frustrations of everyday life. Most repertoires include a considerable amount of humorous material. Cowboy poetry is a popular part of many events that showcase Florida heritage.

Indian River County cattlemen. 1951. Courtesy of Sean Sexton.

There is a story associated with this image of four principals of the Kenmore Cattle company and two cowhands. Kneeling in the foreground is "Mr. Gilbert" Barkoskie, considered to be the grandfather of the Indian River Cattlemen's Association. To his right is Waldo E. Sexton. Standing in back are (left to right) Prescott Gardner, cowhand Perry Savage, Hubert Graves, and Perry's brother, cowhand Charlie Savage.

Later in the 1950s, Perry Savage was gored by a bull in the Moodys' pens out near Twenty Mile Bend while they were working cattle. Someone ran up to him as he lay on the ground and asked, "Did he hurt you Perry?" "No, he kilt me," he replied. They took him to nearby Okeechobee but the hospital was closed, and he died in the vehicle between Okeechobee and Vero Beach.

*Ballad of Bone Mizelle**

by Ruby Leach Carson

At Kissimmee they tell of old M. Bone Mizelle
And the stranger who died on his hands;
How he died in dry season, and that was the reason
He was buried awhile on Bone's lands.

He was buried awhile on that pine and palm isle
In a swamp under Florida's sun
By the Cracker who nursed him, who loved him and cursed him
Just before his demise had begun.

"Jes take this news ka'mly,"Bone wrote to the family.
The deceased had left livin' up north,
"I can send th' remains when there come up some rains
And us pine island folk can go forth.

"So providen' yuh ask it, I'll dig up th' casket". .
Which was done when the season brought rain,
And the river could float the flat-bottomed boat
And the dead boy could travel again.

When Bone went with the coffin, he smiled much too of'en
On the boat and en route to the car.
At the train he said "Gimme one fare from Kissimmee
To Vermont! Ain't this corpse goin' far!"

Thus the money was spent that the family had sent
And a friend of Bone's said the next day:
"So yuh shipped that lad hum?" And Bone said: "No, by gum—
For I thought hit all over this way:

"As his kinfolks air strangers to all of us rangers,
Why not give some dead Cracker this ride?
Why not make all this fuss over some pore ole cuss
Who in life hadn't wallered in pride?

"So instead of that Yank with his money and rank
Who had been 'round and seen lots of fun,
I jes' dug up Bill Redd and I sent him instead
For ole Bill hadn't traveled 'round none."

**Carson used this spelling. The correct spelling is Mizell.*

"Ballad of Bone Mizell," by Ruby Leach Carson. Tinsley, Jim Bob. *Florida Cowhunter: The Life and Times of Bone Mizell,* Gainesville: University Press of Florida, 1990, p. 104.

One of the best known poems about a Florida cow hunter is "The Ballad of Bone Mizelle," written by Ruby Leach Carson in 1939.

Doyle Rigdon. Brighton, September 2008. Photo by Robert L. Stone, courtesy of the State Archives of Florida, image FS060837.

Doyle Rigdon works as cowboy for the Lykes Brothers Ranch. His poems express the excitement and frustrations of the everyday life of the Florida cowboy and are often marked by side-splitting humor.

Rodeo announcer. Kissimmee, February 17, 2008. Photo by Robert L. Stone, courtesy of the State Archives of Florida, image FS060817.

Mounted on a beautiful while horse fitted with a fancy saddle, announcer Jerry Todd welcomes the audience to the 120th Semi-Annual Silver Spurs Rodeo.

Skilled announcers heighten the drama of rodeos by providing colorful, informative commentary. Many entertain audiences with jokes drawn from material polished over the years, some of it received from previous generations. Often they engage in humorous dialog with rodeo clowns. Skilled in the Southern preaching tradition, ministers for "cowboy church" services may also work as rodeo announcers.

Auctioneers play an important role in transacting sales at livestock markets. Although large uniform lots of cattle are increasingly sold through Internet and video auctions, auctioning animals individually remains the best way for many small-scale beef producers to sell cattle. An accomplished auctioneer knows the stock and buyer preferences. He is able to communicate quickly and efficiently, though his rapid patter may be unintelligible to the untrained ear.

Carl Sharp. Kenansville, December 2006. Photo by Robert L. Stone, courtesy of the State Archives of Florida, image FS060831.

Carl Sharp's talent for cowboy poetry performance has made him a popular entertainer at cattlemen association meetings and cowboy gatherings throughout Florida and across the United States.

Painting by Regina Briskey (1945–). Loaned by Jim and Renee Strickland.
Photo by Ray Stanyard.

Cowmen relate stories about unusual events in their working life. This painting depicts an incident in which Jim Strickland rescued his cow-dog from an alligator by lassoing him and pulling him out of a creek in a cypress dome.

Recreation

Rodeo elevates the cowboy's occupational skills from everyday work to a spectator sport that is tremendously popular on and off the ranch. Florida's warm climate accommodates a full calendar of rodeo events year-round. Notable rodeos include Kissimmee's Silver Spurs Rodeo, reputedly the largest in the East, and the Arcadia Rodeo, the oldest in the state.

There are many important roles in the multi-faceted rodeo pageant. Girls and young women compete in horsemanship, appearance, and personality to be selected as rodeo queens. Rodeo clowns perform physical comedy and banter with the emcee. Bullfighters keep fallen riders safe by distracting the bull. Many rodeos include a parade through town before the event. The rodeo opening, or "grand entry," usually features mounted drill teams, rodeo queens and flag-bearers in colorful costumes.

The increasingly popular ranch rodeos demonstrate that working ranches are a vital component of Florida folklife. In these rodeos, ranch-sponsored teams compete in events which better reflect everyday ranch work than conventional rodeos. Some events, such as wild cow milking, may result in humorous chaotic struggles to subdue excited livestock.

More than 200 students compete in Florida High School Rodeo Association (FHSRA) events throughout the state. In addition to traditional women's events such as barrel racing and breakaway roping, girls compete in goat tying, pole bending (a sort of horseback slalom), and team roping. Unlike adult rodeo, which is predominately male, the gender balance in high school rodeo is relatively equal.

Cattle drives reinforce the significance of ranching traditions and inform Floridians about the state's rich ranching history. To celebrate 150 years of statehood in 1995, 600 cattlemen, historians, horse lovers and hearty adventurers drove 1,000 head of native Cracker cattle across Florida. They undertook a similar drive of 500 cattle in 2006 near Kenansville.

"Get off my back." Arcadia, 2008. Photo by Jimmy Peters.

A cowboy attempts to make the eight-second bell in the bareback bronc competition at the 80th Annual Arcadia All-Florida Championship Rodeo.

Little Pig Foot Farm team roping. Williston, March 2008. Photo by Robert L. Stone, courtesy of the State Archives of Florida, image FS060807.

Brandon Jelks (Levy County), who has successfully roped the head of a steer, moves it into position for heeler "Pizza" (Montgomery, Alabama) to rope the hind legs. The Wallace family has been hosting an annual calf- and team-roping event at their Little Pig Foot Farm near Williston since 1996.

Rodeo award belt buckle. Silver and gold. Loaned by Leo Ledesma. Photo by Ray Stanyard.

This buckle was awarded to Cuban-American rodeo champion Leo Ledesma for winning the team roping championship at Davie, Florida in 1998.

Faye Blackstone (1915-2012) executing one of her most famous tricks, a reverse fender drag. Parrish, 1952. Courtesy of the State Archives of Florida, image n036758.

Faye Blackstone achieved national fame as a barrel racer and trick rider, inventing feats such as the daring "back fender drag," which she performed in Wild West shows. Faye's husband, Vick Blackstone (1913-1987), was a rodeo champion. In 1982, he was inducted into the National Rodeo Cowboy Hall of Fame, and Fay into the National Cowgirl Hall of Fame—the only couple to share this honor.

FAYE

Pete Clemons receiving award boots.

Fancy cowboy boots. Leather, sterling silver. Loaned by Pete Clemons. Photo by Robert L. Stone.

These boots were presented to Pete Clemons at the 1955 Silver Spurs Rodeo, Kissimmee, for winning the "Best All Around" title for two consecutive years.

Program booklet, "First Annual Silver Spurs Midwinter Rodeo." Paper, 8.75 inches x 11.75 inches, 20 pages. Event dates: February 25-26, 1951. Courtesy of Silver Spurs Rodeo. Photo by Ray Stanyard.

The Silver Spurs Rodeo is held each February and June in Kissimmee. It is the largest rodeo east of the Mississippi and is annually ranked among the top 50 events sanctioned by the Professional Rodeo Cowboys Association (PRCA). The program booklet cover depicts the old rodeo grounds. Today the event is held in an indoor climate-controlled 8,300-seat arena.

Pete Clemons with Buster Kenton. Kissimmee, 1949. Courtesy of Pete Clemons.

When Pete Clemons was asked to represent the Kissimmee Jaycees at a national convention rodeo in Colorado Springs in 1949, they fixed up a pickup truck to look like a chuck wagon. Buster Kenton (right) decorated it with a cartoon character he called Cowboy Jake. Clemons won the Best All Around competition to the dismay of the western competitors. Cowboy Jake became the mascot for Osceola High School.

Clemons was tremendously successful in his 25-year rodeo career. He won the Best All Around Cowboy title at the Silver Spurs Rodeo in Kissimmee nine times. This required excelling in five events: saddle bronc, bareback bronc, calf roping, steer wrestling, and bull riding. He competed throughout the U.S., as well as in Canada, Cuba, and South America.

Rodeo award belt buckle. Silver and gold. Loaned by Pete Clemons. Photo by Robert L. Stone.

This buckle, inscribed "All Around Cowboy, San Antonio, Florida, 1951," was presented to Pete Clemons.

Rodeo chaps. Leather. Loaned by Matt Condo.
Photo by Ray Stanyard.

These bat-wing style rodeo chaps were made from the hide of a mule deer killed by Florida rodeo champion Matt Condo when he was competing in Colorado ca. 1961.

Pat Hansel presents a buckle for Best All Around Cowboy to Matt Condo at the Indiantown Rodeo. Photo by Charles Covington, courtesy of Debbie Hansel Carlton and Leola Parker Hansel.

Matt Condo, bronc rider. Palatka, 1971.
Courtesy of Matt Condo.

Matt Condo rides a bronco named Claim Jumper at the 1971 Florida Rodeo Cowboys Finals Rodeo, Palatka. Condo garnered Best All-Around Cowboy honors at the Arcadia Rodeo five times. Seminole rodeo legend Josiah Johns, who is next to ride, is in the chute, wearing a light-colored hat.

Josiah Johns, Seminole rodeo champion. Las Vegas, 1976.
Photo by James Fain, courtesy of Willie Johns.

Josiah Johns ropes a calf at the 1976 Indian National Finals Rodeo in Las Vegas, Nevada. The legendary Seminole cowboy garnered Best All Around Cowboy honors five times at the Arcadia All-Florida Championship and Chalo Nitka rodeos and was three-time steer wrestling champion of the Florida Cowboy Association.

Bareback bronc rigging, leather with steel D-rings. Loaned by Matt Condo. Photo by Ray Stanyard.

Made by Bob Fraker, Ft. Collins, Colorado, for Matt Condo, ca. 1965. The rigging is secured to the horse with a cinch.

Pat Hansel

The All Florida Championship Rodeo, held in Arcadia since 1928, is Florida's oldest rodeo and the centerpiece of the state's vibrant rodeo scene. Pat Hansel was known in Arcadia as "Mr. Rodeo."

As a stock contractor Pat provided some of the best roughstock—bucking bulls and broncos—of his era. His bulls included Wooley Booger, a big black Brahman-Angus crossbreed; Bearcat, a quick, compact red bull; and the legendary Seminole Chief. Seminole Chief so dazzled the audience, contestants and stockmen with his performances at the 1981 PRCA National Finals Rodeo in Oklahoma City that he brought a $20,000 selling price, an astronomical sum at that time.

Pat's bucking horses were always among the best, too. Claim Jumper was the most renowned, having a rodeo career that spanned eighteen years.

Pat is remembered by thousands for his warm and colorful commentary as rodeo announcer at the Arcadia arena. His close friend, livestock auctioneer Wendell Cooper, was inspired by Pat's friendship and contributions to the rodeo community to write this poem for Pat's funeral.

Photographs and artifacts loaned by Debbie Hansel Carlton and Leola Parker Hansel.

41st Arcadia All Florida-Championship Rodeo Rodeo program booklet, 1969. Jacki Duncan Condo (L) and Debbie Hansel Carlton pose for a Coke advertisement.

Pat Hansel announcing the rodeo in Indiantown, 1972.

Rodeo Memories of Pat

Pat Hansel was a man you needed to know
If you were a Florida Cowboy and loved rodeo.
He knew all the contestants and could call 'em by name.
Pat changed a lot of things in the rodeo game.

In the old days rodeos weren't organized and the judging was rough,
The stock wasn't even and winning was tough.
There was a lot of complaining, most times about pay.
Pat set out to change all that, said he would find a way.

He was one of the first in Florida to join the PRCA.
Pat's efforts in rodeo are still remembered today,
He took a bull to the national finals because he could.
That bull brought $20 thousand, no one thought he would.

Now things were better and it wasn't just luck.
Pat raised his own stock and those suckers could buck.
His bulls were his pride and their names would inspire.
A lot of bull riders still remember Ring of Fire.

Another little bull that was seldom rode,
A full-blooded Angus that was a sight to behold.
A little guy that would spin right outside the gate,
he was so small Pat just called him six-and-seven-eights

How he came up with those names sure puzzles me.
He must have stayed up nights thinking what the next one would be
Those names were sure different I remember one now,
The funniest of them all was one he called screech owl

His friends will sure miss him of this there's no doubt.
He taught all us cowboys what rodeo was all about.
Be fair and honest, try hard as you can
In rodeo and in life, you will be a top hand

I'm glad I could count Pat a personal friend of mine.
If you ever needed help, he'd be there every time.
If you were down and out Pat would lend a hand,
But he'd make you get up saying, "Son act like a man."

Pat was a winner right up to the end,
surrounded by the love of his family and the respect of his friends
His last win was his best; he's left this earthly sod.
Pat went home to be with Jesus and the family of God.

—Wendell Cooper March 24th, 2007

A note from Pat Hansel's daughter, Debbie Hansel Carlton:
Wendell Cooper was fighting a battle with cancer when he wrote this wonderful poem. He lost that battle and also went home to be with Jesus and the family of God a year later. Thank you God for making our lives richer because of Daddy Pat and Wendell.

Custom-made belt buckle worn by Pat Hansel, silver and gold. The figure of Pat was originally holding a microphone, which has broken off.

Pat Hansel (L) and unidentified sign painter with Pat's rodeo stock truck, Arcadia, 1966. Photo by George Shupka.

Rodeo clown Bill Shaw fends off Pat Hansel's bucking bull, Seminole Chief, at the National Finals Rodeo, Oklahoma City, Oklahoma, 1981. Pat Hansel is in two-tone western jacket, center. Photo by James Fain.

Red checkerboard cowboy boots made for Pat Hansel by the Tony Lama Company. Pat had several pairs of boots custom made for his rodeo announcer attire.

Tater Porter, bull riding champion. 2007. Courtesy of Enterprise Rent-A-Car.

Tater Porter of Kenansville won the Professional Bull Riders, Inc. (PBR) World Finals in Las Vegas in 2000. He retired in 2007, but continues to do public relations work for PBR. When not traveling, he works with the cowboy crew at the VIDA Ranch in Osceola County.

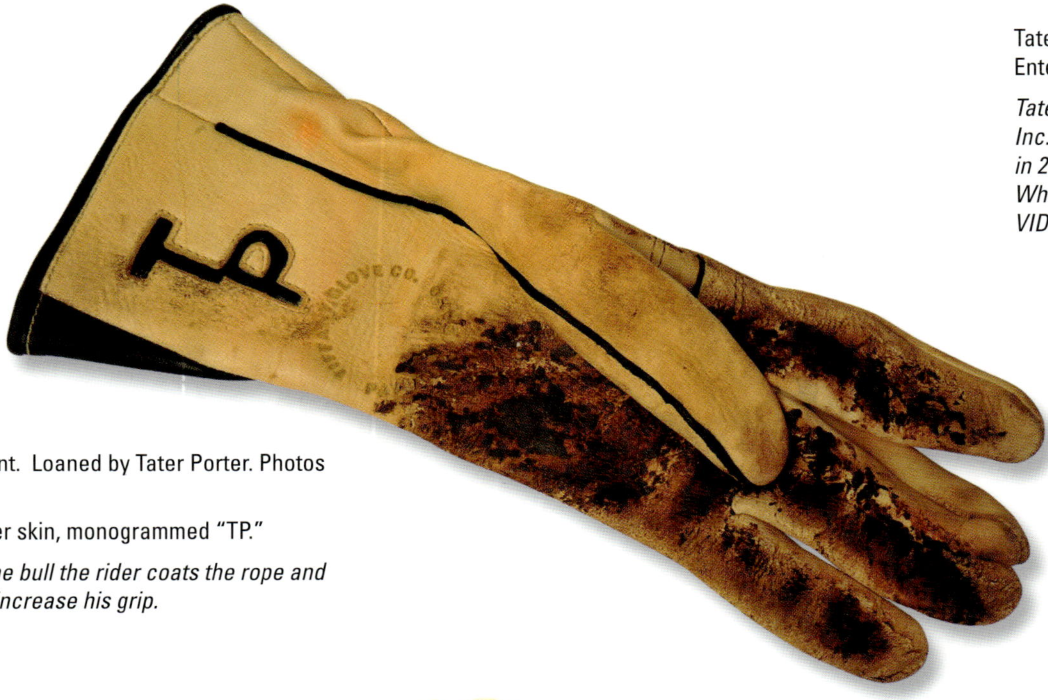

Bull riding equipment. Loaned by Tater Porter. Photos by Ray Stanyard.

Bull rider glove, deer skin, monogrammed “TP.”

Prior to mounting the bull the rider coats the rope and glove with rosin to increase his grip.

Bull rider spurs. Steel with silver overlay “TP” monogram.

Spurs help the rider grip the bucking bull. A bull’s hide is very thick and the spurs do not hurt it. The spur shanks are angled inward for a better riding position and the specially shaped rowels are locked to prevent them from rotating.

Bull rope, nylon.

The flat braided rope is wrapped around the bull's chest just behind the front legs. The tail of the rope is threaded through the loop on the opposite end and tightened around the bull. The rider grips the handle then wraps the tail around his hand to further secure his grip.

Bull bell, steel.

The weight of the heavy bell acts to release the rope from the bull once the rider has dismounted. This bell was crushed when a bull stepped on it.

Pole bending. Williston, April 2007. Photo by Robert L. Stone, courtesy of the State Archives of Florida, image FS060834.

A competitor negotiates a tight turn in the girls' pole bending competition at the annual FHSRA rodeo

Grand Entry. Williston, April 2007. Photo by Robert L. Stone, courtesy of the Florida Folklife Program/ Florida Department of State.

Carrying the American flag, Ashley Mills leads girls on horseback bearing colorful flags to create a dramatic "grand entry" at the annual FHSRA rodeo, Williston Horseman's Park.

Boot and spur of bull rider Stephen Keighley, Williston, 2007. Photo by Robert L. Stone, courtesy of the State Archives of Florida, image FS060808.

Bull riding spurs are fastened tightly to the boot with a padded leather strap and held firmly with a twisted wire wrapped under the boot, in front of the heel. A leather thong wrapped tightly around the boot top secures the boot to the foot.

High School Rodeo bull riding. Williston, April 2007. Photo by Robert L. Stone, courtesy of the State Archives of Florida, image FS060824.

Robert Carter of Cross City. Some high school bull riders earn scholarships to compete on college rodeo teams, others go on to compete in professional events.

Wild Cow Milking, Polk County Ranch Rodeo. Bartow, March 2008. Photo by Robert L. Stone, courtesy of the Florida Folklife Program/Florida Department of State.

Chaos ensues as (left to right) the Overstreet (blue neckerchiefs), Lake Hancock (green), and D.R. Fussell (khaki) ranch teams compete in the wild cow milking event. The object is to obtain a small amount of milk in a plastic water bottle and run with it to the judge standing on the sideline.

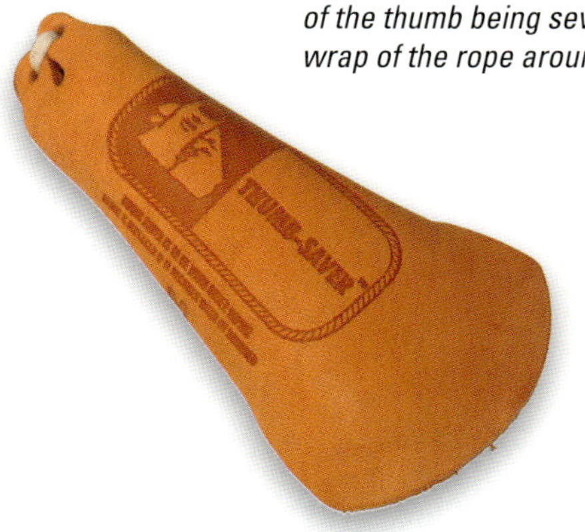

Thumb-Saver. Leather. Donated by Billy Malphurs. Photo by Ray Stanyard.

Some competitive ropers wear a leather guard under the glove to prevent the rather common injury of the thumb being severed when caught in the dally wrap of the rope around the saddle horn.

Clown on Seahorse. Williston, May 2008. Photo by Robert L. Stone, courtesy of the Florida Folklife Program/Florida Department of State.

Chad "Cracker" Johnson is a part-time rodeo clown from Chiefland. He engages in humorous banter with the announcer by means of a cordless hands-free microphone.

Okeechobee Ranch Rodeo. July 2007. Photo by Robert L. Stone, courtesy of the State Archives of Florida, image FS060846.

Alan Lanier, center, and teammates work quickly in the team doctoring competition at the ranch rodeo that was part of the 2007 National Day of the American Cowboy in Okeechobee. Each team had to rope a cow, tie it and paint it with a marker, with the fastest team declared the winner.

Calf Branding, Polk County Ranch Rodeo, Bartow, March 2008. Photo by Robert L. Stone, courtesy of the Florida Folklife Program/Florida Department of State.

Megan Talbo runs to the finish line in the calf branding competition at the Polk County Ranch Rodeo. Many ranch rodeos require each team to include at least one woman.

Great Florida Cattle Drive of '06. Osceola County, December 2006.
Photo by Carlton Ward, Jr.

More than 500 people participated in the four-day, 50-mile cattle drive.

Rodeo Queens. Kissimmee, February 2008. Photo by Robert L. Stone, courtesy of the Florida Folklife Program/Florida Department of State.

Left to right: Amy Wilson, Miss Rodeo America; Ryleigh Tyson, Little Miss Silver Spurs; and Samantha Roberts, Miss Silver Spurs, at the Silver Spurs Arena

Seminole Rodeo Queens. Brighton Seminole Indian Reservation, November 2006. Photo by William Mansfield, courtesy of the Florida Folklife Program/Florida Department of State.

Left to right: Le Anne Billie, Seminole Rodeo Queen; Heather Peterson, Junior Seminole Rodeo Queen; and Jewel Buck, Brighton Rodeo Princess.

Ricky Doctor. Seminole Rodeo Queen beaded crown. Glass, cotton thread, fabric. Big Cypress Seminole Indian Reservation, 1993. Loaned by Joleigh Johns. Photo by Ray Stanyard.

Joleigh Johns was crowned Seminole Rodeo Queen in 1993. Ricky Doctor fashioned this crown, which features the traditional Seminole medicine colors red, black, yellow and white. He learned his craft from his mother, Annie Doctor.

The Future

Cattle Ranching in the Twenty-first Century

In January 2008, there were over 1.7 million cattle in Florida, including nearly 1 million head of beef cows. Florida was 12th in the nation for beef cattle and 18th for total cattle and calves. Industry officials assess the annual economic impact of beef cattle ranches at nearly $4 billion.

To maintain the integrity of their land and herds, Florida ranchers have always been environmental stewards. They constantly search for ways to protect pastures from overgrazing, maintain a permanent water supply, improve forage, and control insects and disease. Ranchers play an important role in the preservation of the diverse wildlife and natural habitats found on ranches. To ensure that ranch lands are not consumed by development, owners increasingly enter into conservation easements and other legal arrangements that ensure long term stewardship.

The efforts of dedicated individuals and organizations have ensured the future of Cracker cattle and horses. In 1970, Florida's Department of Agriculture and Consumer Services began working with the Florida Cattlemen's Association to preserve Cracker cattle. Formed in 1988, the Florida Cracker Cattle Association and Florida Cracker Horse Association adopted breed standards and appointed evaluation committees to select animals to be registered as foundation stock. Although they lack broad-based commercial value, Cracker cattle are prized as tangible links to the state's rich ranching history. Cracker horses are widely popular because they can be trained for all riding disciplines and make excellent ranch horses.

Several organizations provide young people with training and experience. In addition to vocational agriculture courses in middle and high schools, FFA (formerly Future Farmers of America) and 4-H offer extra-curricular activities that teach the business of raising cattle and develop leadership skills. Industry organizations provide programs such as Junior Cattlemen, the Cattlemen's Sweetheart, and Beef Ambassadors, while the Seminole Tribe has summer co-op programs for interested students.

Despite many challenges, the future of Florida cattle ranching is secure because it is more than just a business; it is a cultural heritage and a way of life.

Tom Everett, Sr. with cell phone. Sumter County, September 2006. Photo by Robert L. Stone, courtesy of the Florida Folklife Program/Florida Department of State.

The cell phone has become an indispensable means of communication for cattlemen. Ranch life is shaped by changing technology. Research continues to improve fertility, breeding, pasture management and other factors that result in more efficient beef production.

Justin Gopher with Cracker horse. Big Cypress Seminole Indian Reservation, July 2007.
Photo by Robert L. Stone, courtesy of the State Archives of Florida, image FS060827.

Justin Gopher takes a break during the annual roundup. The small, agile horses descended from stock introduced by the Spanish in the 16th century make exceptional ranch horses. Seminoles and non-Indians have used them for centuries. In 2008, the Florida legislature declared the Florida Cracker Horse the state's official heritage horse.

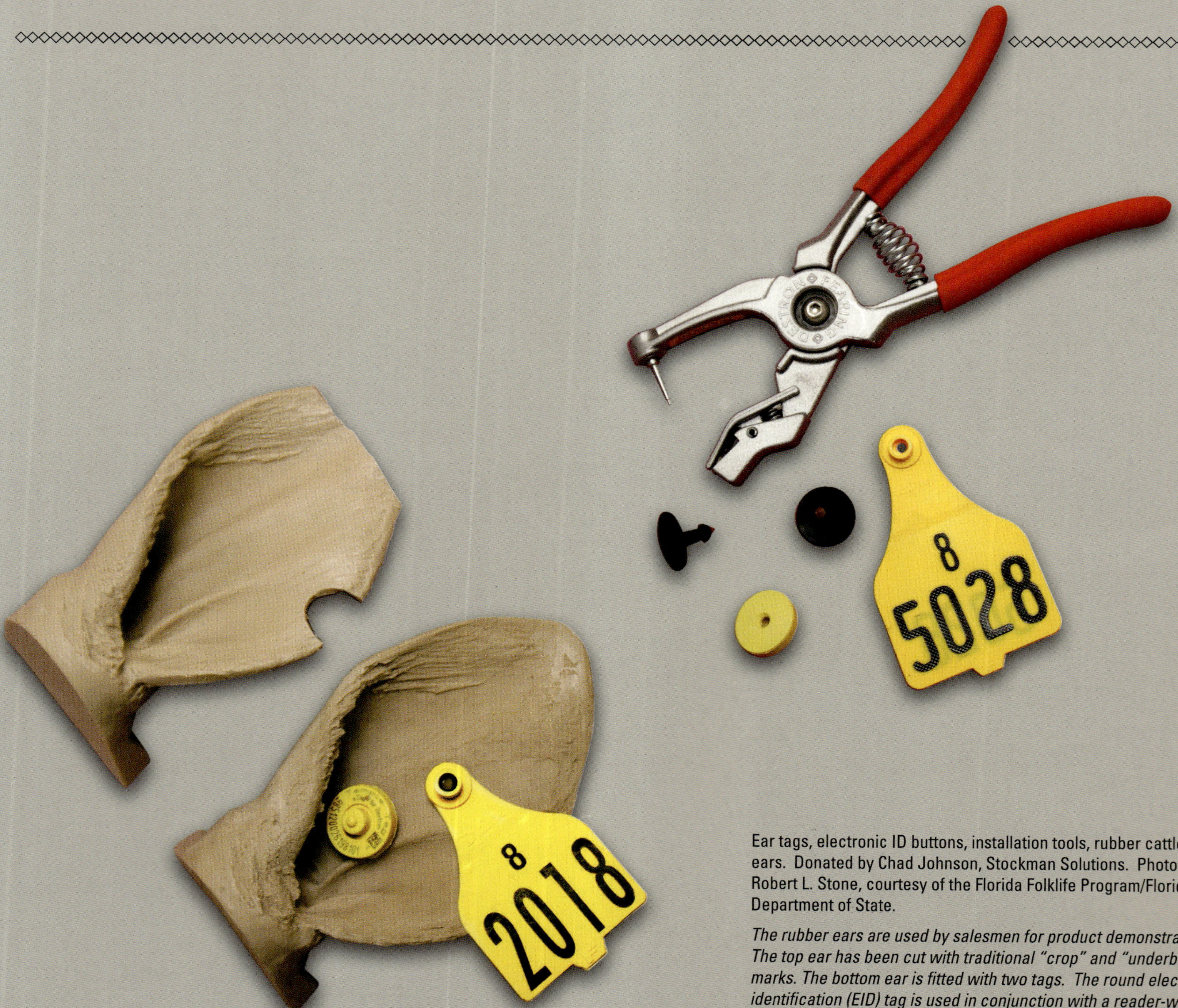

Ear tags, electronic ID buttons, installation tools, rubber cattle ears. Donated by Chad Johnson, Stockman Solutions. Photo by Robert L. Stone, courtesy of the Florida Folklife Program/Florida Department of State.

The rubber ears are used by salesmen for product demonstration. The top ear has been cut with traditional "crop" and "underbit" marks. The bottom ear is fitted with two tags. The round electronic identification (EID) tag is used in conjunction with a reader-wand and computer software to track breeding, nutrition, weight, vaccination and other aspects of an animal's history. The large numbered tag is for visual identification. Special pliers are used to fasten the black self-piercing studs to the tags.

Hoof trimming lesson. Okeechobee, September 2008. Photo by Robert L. Stone, courtesy of the Florida Folklife Program/Florida Department of State.

Okeechobee High School students Kayla Davis and Matthew Dorriety watch closely as Agricultural Science teacher Roger McWaters shows them how to trim and dress a horse's hoof.

Leoma Simmons at laptop computer. Big Cypress Seminole Indian Reservation, July 2007. Photo by Robert L. Stone, courtesy of the Florida Folklife Program/Florida Department of State.

The Seminole Tribe of Florida uses the latest technology in their cattle operation. All their cattle are fitted with electronic ID tags, which track all aspects of the animal's history. Cattle are weighed on digital scales, marketed by video auction, and tracked using sophisticated software.

Kaley Dees with cow. Okeechobee, September 2008. Photo by Robert L. Stone, courtesy of the Florida Folklife Program/Florida Department of State.

Okeechobee High School junior Kaley Dees at her family ranch with the registered Angus heifer she shows in FFA and 4-H events. She aspires to become a large animal veterinarian.

Florida Cowboy Glossary

- Cabbage or cabbage tree: a cabbage or sabal palm, the Florida state tree.
- Cold-backed horse: balks at being saddled or mounted and may buck.
- Crevice: a holding area adjacent to a holding pen.
- Brammer or Bremmer: common pronunciation of Brahma or Brahman cattle and their derivatives.
- Cowhunter: a cowman who hunted cattle on horseback, usually with dogs, and more recently referring to one participating in a round-up. The cowhunter was the same as a cowboy.
- Dally: to tie a roped animal to the saddle horn with one or more wraps of the rope. This enables the rope to be reeled in or out.
- Dobie: an orphaned calf.
- Gyp: a female dog.
- Hammock: mixed hardwood and palm islands on ranches providing watersheds, grazing areas, and wildlife habitats
- Hard-and-fast tie: to fasten a roped animal to the saddle horn with a knot or loop in a manner which does not allow the rope to be played out. A practice favored by many Florida cowboys.
- Heifer: a female calf and young female cow.
- Mammy-up: when a cow and calf are re-joined after separating. The ability to mammy-up cows and calves is a valued skill among cowboys.
- Rank: aggressive, mean, ill-tempered.
- Ride out: to ride into a herd and separate out cows or cow/calf pairs.
- Steer: castrated male cattle raised for beef.

Jon Kral

The following eight black and white photos are the work of Pulitzer Prize winning photojournalist Jon Kral. From 1975-2000 he documented Florida cowboys in south-central and central Florida at work on several ranches and at home with their families. In 2012, the South Florida State College Museum of Florida Art and Culture (SFSC MOFAC) acquired Kral's Cracker Cowboy Collection of 109 photographic prints. The SFSC MOFAC is housed at the SFSC Highlands Campus in Avon Park, Florida. For more information, call 863-784-7240 or visit www.mofac.org.

Hughes home, Kenansville, 1995.

John Davis, Alico Ranch, LaBelle, 1995.

Mess Hall, Alico Ranch, LaBelle, 1995.

Lightsey Cattle Company, Lake Kissimmee, 1995.

Leslie Tyson, Kenansville, 1990.

Junior Mills, High Horse Ranch, Indiantown, 1990.

Trail Boss Jennings Overstreet and crew, Florida Roundup, 1995.

Geech Partin, 75, Kissimmee, 1989.

Acknowledgements and Credits

The Exhibition

The exhibition presented in this catalog was produced by the Florida Folklife Program, Division of Historical Resources, Florida Department of State, and Florida Cultural Resources, Inc. Major funding was provided by the Florida Cattlemen's Association, the Florida Cattlemen's Foundation, the National Endowment for the Arts, the Florida Humanities Council, the National Endowment for the Humanities, the Florida Division of Cultural Affairs, and the Florida Cracker Cattle Association.

Robert L. Stone and Tina Bucuvalas curated the exhibition and wrote the text. Little Exhibit & Graphic Design created the exhibit. Project Advisors included Bud Adams, Joe A. Akerman, Anne McCudden, Dana Ste. Claire, Renee Strickland, and Stephen Stuempfle. We are grateful for the consistent editorial and conceptual support of Susanne Hunt, /Bureau of Historic Preservation. Thanks also to the assistance of Museum of Florida History staff 's Jeana Brunson, Charity Wood and Kieran Orr for their assistance in artifact acquisition, exhibit design, and coordination of exhibition loans and travel.

Thanks to the 144 donors statewide who contributed so that their ranch brands, many of which are family heirlooms, could be displayed. Special thanks for the generous donations made by Doyle Carlton III, Woody and Grace Larson, J. B. Kittrell, the Putnam family, Lalla Rook Tomkins, Iris Wall, and Imogene Yarborough.

The Exhibition Catalog

Editor: Robert L. Stone
Graphic Design: DECA Design Inc., Gainesville, Florida
Printing: StorterChilds Printing Co. Inc., Gainesville, Florida

Exhibit Catalog Committee: Susannah Booth, Jim Strickland, Ned Waters and Frank Yong

We are grateful to the guest photographers for granting use of their images:

Carlton Ward, Jr. – www.carltonward.com
Jon Kral – www.kralphoto.com
Jim Peters – www.jimpetersphotography.com
Bob Montanaro – www.floridacattleranch.org

There are many people to thank for their generous contributions of information, expertise, images, and assistance:

Frank Batey
Pete and Susanne Clemons
Destron Fearing Co.
Douglas B. Conner, Connerton, L.L.C.
Doyle Conner, Jr.
Donita Dampier, Osceola County Historical Society
Florida Cattlemen's Association staff
Florida Department of Agriculture and Consumer Services
Florida State Fair Authority
Jack Gillen
Robert and Emily Hobby
Alex Johns
Willie Johns
Chad Johnson
William Mansfield
Stephen Monroe
Sean Sexton
Stockman Solutions
Judy Trotta
Ned Waters
Mike Wilder
Imogene Yarborough
Bobby Yates

Whaley Cattle Co.

Peace Valley Ranch

SS Farms, LLC

Carlton Ranch

Weeks Cattle Company

Wynn & Sons Cattle

Strayhorn's Double Spade Ranch

Ryals Cattle Co.

C.J. Cattle Co., LLP

Yellow Pine Farm, Inc.

Larry E. Rooks

Valentine M. Rooks

Rowell Ranch

Rollins Ranches, LLC

Dee Wooten

Will McClelland

Ira McClelland

Litard Knot Ranch

Turner Cattle Company

JB Ranch

Kissimmee Park Properties, LLC

Marlin W. Hilliard

Joe M. Hilliard

Joe A. Hilliard

Ramsey's Cattle Co.

A1 Bar Ranch, Inc.

Tousaint & Strickland

Simon Grove & Cattle Co., Inc.

Stardust Angus Ranch

Tall Timber Cattle & Grove, Inc.

Strickland Ranch

Old River Cattle Company

Nancy Pearce Phares

Lazy JP Ranch

Justin Rooks

Highland Farms, Inc.

Lucky 2D Ranch

Mound Creek Ranch

O'Dell Properties

Bar LH Cattle Co.

C2 Ranch, Inc.

Gilbert A. & Ruth Tucker

Parchman Prairie Ranch

Quincey Cattle Company

Rockin H Ranch

Hancock Holding Company

Hayes Land & Cattle Company

Hendrick Cattle Company

Rocking J E Ranch

Rodgers Ranch

Rolling R Ranch

High Horse Ranch

Jacob & Danielle Larson

Jumper Creek Ranch

Sam Sikes

Seventh Heaven

Simmons Farm

Larson Dairy, Inc.

Lightsey Cattle Co.

Little Big Horn

Smith Brothers

South Branch Ranch

Square C Ranch

Barto Farms

Bar L Cattle

7 L Bar Ranch

Handley Ranch

Greg & Janet Corson Ranch

Edward L. Partin Ranch, Inc.

Dietrich's Flying D Ranch

Crawford Ranch

Crawford Ranch